Acrylate Polymers for Advanced Applications

Edited by Ángel Serrano-Aroca and Sanjukta Deb

Published in London, United Kingdom

IntechOpen

Acrylate Polymers for Advanced Applications
http://dx.doi.org/10.5772/intechopen.77563
Edited by Ángel Serrano-Aroca and Sanjukta Deb

Contributors
Thomas Swift, Ramesh Rudrapati, Kingsley Kema Ajekwene, Ángel Serrano-Aroca, Sanjukta Deb

First published in London, United Kingdom, 2020 by IntechOpen
IntechOpen is the global imprint of INTECHOPEN LIMITED, registered in England and Wales, registration number: 11086078, 7th floor, 10 Lower Thames Street, London, EC3R 6AF, United Kingdom
Printed in Croatia

British Library Cataloguing-in-Publication Data
A catalogue record for this book is available from the British Library

Additional hard and PDF copies can be obtained from orders@intechopen.com

Acrylate Polymers for Advanced Applications
Edited by Ángel Serrano-Aroca and Sanjukta Deb
p. cm.
Print ISBN 978-1-78985-183-0
Online ISBN 978-1-78985-184-7
eBook (PDF) ISBN 978-1-78984-711-6

For EU product safety concerns:
IN TECH d.o.o., Prolaz Marije Krucifikse Kozulić 3, 51000 Rijeka, Croatia, info@intechopen.com or visit our website at intechopen.com.

Supporting open minds since 2005

Meet the editors

Professor Serrano-Aroca holds a PhD in Chemical Engineering and currently teaches bioengineering at the Universidad Católica de Valencia San Vicente Mártir. His research interest is developing medical materials and devices for advanced applications such as antimicrobial therapy, tissue engineering, wound healing, etc. He is currently Vice Dean of Biotechnology and Principal Investigator of the Biomaterials and Bioengineering Lab at the Centro de Investigación Tranlacional San Alberto Magno.

Professor Sanjukta Deb is a professor in biomaterials science at King's College London. The main theme of her research is developing innovative biomaterials and biomimetic scaffolds to restore function of traumatized/diseased tissue for clinical translation. She is currently the Chair of the Royal Society of Chemistry: Biomaterials Chemistry interest group and the ex-President of the UK Society of Biomaterials.

Contents

Preface

The field of polymer science and engineering includes research in multiple disciplines including chemistry, physics, engineering, biotechnology and medicine, among many others. Polymers include synthetic polymers such as plastics and elastomers, and natural biopolymers. Acrylates are synthetic polymers classified as thermoplastic resins used in a great number of industrial products, ranging from bone cement or contact lenses in biomedical applications to artificial nails, diapers, cosmetics, orthopaedics, paints and coatings, adhesives and textiles in other industries. Acrylates are made from acrylate monomers that contain vinyl groups and possess a wide range of properties ranging from super-absorbency, transparency, flexibility, toughness and hardness. Currently, there are many polymers produced from acrylate monomers. Here we analyse how their solution properties are pH dependent and the effect of the state of ionisation affects their solution properties. Acrylate polymers such as poly (acrylic acid) and poly (methacrylic acid) are polyelectrolytes, with ionisable functional groups that render them stimuli responsive, changing their hydrodynamic volume. Poly (acrylamide) is a mass-produced acrylate material utilised in a variety of industrial applications, often produced from an anionic and cationic co-monomer. Poly (N-isopropyl acrylamide) is a thermally responsive acrylate material with applications in smart bioengineering. Other advanced applications have been developed with polycarbonate antiballistic and acrylic heat resistant glass due to their important properties such high impact strength, transparency, reflective index and chemical reactivity. In the field of biomedicine, acrylic-based materials have been used over many years due to their versatile properties. In fact, many different acrylate polymers have been approved by the US Food and Drug Administration (FDA) for biomedical use in humans. Thus, they are frequently used in orthopaedics, ophthalmologic devices, antimicrobial therapy, tissue engineering applications and dental applications. Many sophisticated methods and techniques have been developed in the last few decades in order to expand their potential applications in the biomedical industry. Thus, successful scientific enhancements have been achieved with respect to mechanical performance, electrical, thermal properties, diffusion, biological behaviour, antimicrobial activity and porosity. Multicomponent acrylate-based polymeric platforms have been developed as interpenetrating polymer networks or in combination with other materials such as fibres, nanofibres, carbon nanomaterials and its derivatives and/or other different types of nanoparticles in the form of composite or nanocomposite biomaterials. Furthermore, in bioengineering, acrylic porous supports (scaffolds) need to be synthesised with the necessary degree, type and morphology of pores using advanced technological fabrication techniques.

This book presents five chapters in two sections on recent information about acrylate polymers paying more attention to their properties and advanced applications.

The first section presents three chapters dealing with the properties and applications of acrylate-based materials. The first chapter focuses on four commonly-used examples of acrylate polymers and studies how their solution properties are pH dependent and how their state of ionisation can affect their solution properties. The second chapter deals with polymer material classification as acrylic heat resistant glass and polycarbonate antiballistic glass. The third chapter highlights the characteristic properties and applications of acrylates, its derivatives and copolymers.

The second section of this book contains two chapters on acrylate-based materials in biomedicine and bioengineering. The first chapter of this section presents a review of acrylic-based materials used in biomedical and bioengineering applications and the strategies developed so far to enhance their physical, chemical and biological properties. The second chapter focuses on acrylic-based hydrogels as biomaterials. This book provides information about acrylate polymers that we expect to be very useful for researchers working in this exciting area.

Dr. Ángel Serrano-Aroca
Professor,
Biomaterials and Bioengineering Lab,
Centro de Investigación Traslacional San Alberto Magno,
Universidad Católica de Valencia San Vicente Mártir,
Spain

Dr. Sanjukta Deb
King's College London,
UK

Section 1

Acrylate Polymers: Properties and Applications

Chapter 1

pH Dependence of Acrylate-Derivative Polyelectrolyte Properties

Thomas Swift

Abstract

There are many polymers formed of acrylate monomers in existence. Here we interrogate four commonly-used examples and study how their solution properties are pH dependent, or how their state of ionisation can affect their solution properties. Poly(acrylic acid) and poly(methacrylic acid) are both polyelectrolytes, with ionisable functional groups that make them stimuli responsive, changing their hydrodynamic volume. Poly(acrylamide) is a mass-produced material used in a variety of industrial applications, often with an anionic and cationic co-monomer, which dictates both its efficacy and impact on the environment. Poly(*N*-isopropyl acrylamide) is a thermally responsive material with applications in smart bioengineering. In solution, these materials can interact with each other due to competing hydrogen bonding interactions. However, this interpolymer complexation is dependent on both the ionisation, and the conformational state, of the polymers involved. This review focuses on the results from fluorescence tagging and turbidimetric techniques.

Keywords: poly(acrylic acid), poly(methacrylic acid), poly(acrylamide), poly(*N*-isopropylacrylamide), stimuli responsive, interpolymer complexation, hydrodynamic volume, solution properties

1. Introduction

A common feature of the many polymer systems formed from acrylate monomers is their hydrophilicity; apparent either from their increased absorbency, wettability or increased solubility. Whilst the latter is often overlooked in materials science, it is of vital importance to a range of industries, as a multitude of polyacrylates form vital components in commercial products too varied to list, but including dispersants, adhesives, emulsifiers, lubricants, flocculants, thickeners, surfactants, sensors, delivery agents, coatings, chromatographic phases, grouting, passivation and many more. As of 2018, the multi-million tonne polyacrylate global market is still rising with an annual growth greater than 6% [1]. Research over the last 20 years into controlled radical polymerisation, and copolymerisation, has provided increased insight into the distinct properties of these materials. However, even 50 years after the initial patenting of poly(acrylic acid) [2], new discoveries about its fundamental properties are still being made [3].

In solution many, but not all, acrylate copolymers act as polyelectrolytes, containing ionisable repeat units; and thus show some form of stimuli-response to pH.

The solution forces that govern these properties are the same that give function to biological macromolecules (i.e., peptides, proteins, DNA) and so many polyelectrolytes have been used as simple models for these more complex systems. However, due to their applications are so widespread and varied, it is essential to any chemist or engineer working with these sensitive materials to acquire some understanding of the need to control their pH.

Depending on the nature of these ionisable repeat units, a polymer can be classified as a 'weak' or 'strong' polyelectrolyte, governed by the pK_a of the ionisable groups. As samples containing carboxylic acid repeat units dissociate relatively easily, they fall into the former category. The chemical structure of ionisation (or dissociation/neutralisation) is thus:

$$RCOOH \leftrightharpoons RCOO^- + H^+ \tag{1}$$

and the dissociation constant (α) can be described by the Henderson-Hasselbalch equation

$$\alpha = \left([X] + [H^+] - [OH^-]\right)/[RCOOH] \tag{2}$$

$$pH = pK_a + \log\{\alpha/(1 - \alpha)\} \tag{3}$$

where X is the ionising (titrating) species and pK_a the dissociation constant; the pH at which 50% of the carboxylic groups have been ionised. However, for a polyacid, this is a more contentious issue than studying small molecules due to each acid group is affected by the presence of neighbouring repeat units, which thus modify their titration behaviour. In general, the first COOH group on a polymer backbone shows a similar pK_a to a small molecule analogue. However, as the polymer chain becomes increasingly ionised, the building negative charge constrains further deprotonation, and the pK_a value alters with increasing pH. In this behaviour, particularly polymeric electrolytes show divergent behaviour from small molecules, and Katchalksy and Spitnik proposed a revision to the Henderson Hasselbalch Equation [4].

$$pH = pK_a + n \log\{\alpha/(1 - \alpha)\} \tag{4}$$

where n is a constant dependent on the ionic strength of the solution and the strength of the polyacid. In a stationary solution, this plot should produce a straight line (slope n, intercept pK_a). However, this is rarely observed, particularly in aqueous solutions, and this was the first indication researchers had that many polymeric macromolecules undergo a conformational rearrangement on the nanoscale in response to chemical ionization [4, 5]. Over the years, this has proven fertile ground for research, with poly(carboxylic acid)s receiving particular attention in the literature as they are excellent, chemically distinct, model systems [3, 4, 5–13]. However, even non-responsive systems, such as poly(acrylamide), have been found to demonstrate responsible macromolecular behaviour in the presence of corresponding polymer systems *via* a process of interpolymer complex formation [14]. Many polyacrylates engage in hydrogen bond driven complex interactions. The field has proven to be extremely complex due to the multitude of competing factors that affect this often weak, almost always labile, interface.

This chapter will discuss recent advances in the study of pH dependent polyacrylate solution behaviour, examining our improvements in understanding of weak polyelectrolyte systems. Critically this review limits itself to studies of linear polymer systems, as the properties of branched, or crosslinked, macromolecules are fundamentally different [15, 16] and warrant further, separate discussion.

2. Poly(carboxylic acids)

The two most comprehensively studied synthetic poly(carboxylic acid)s within the literature are poly(acrylic acid) (PAA) and poly(methacrylic acid) (PMAA) respectively. Both contain a carboxylic acid repeat unit that dissociated to form a negatively charged anion in low pH aqueous solutions. The additional methyl group on the methacrylic acid functional group gives PMAA a degree of amphiphilic behaviour [17] depending on the degree of ionisation (**Figure 1**).

This additional hydrophobicity dominates the solution properties of PMAA, leading to the aforementioned 'anomalous' Henderson Hasselbalch titration behaviour [4, 5, 7, 9], whilst PAA has long been considered a more 'ideal' system [18] as it does not undergo as dramatic a macroscopic switch. As the carboxylic acid group can only be classed as hydrophilic when the functional monomer is protonated, PMAA undergoes a rapid swelling as the pH is increased, becoming an entirely hydrophobic material with increasing anionic charge along the backbone. Extensive investigations have been carried out into its behaviour using diverse methods and techniques: pontentiometry [4, 5, 7, 10, 19], viscometry [8, 11], Raman spectroscopy [20], scattering methods [21–23] and fluorescence probe interrogation techniques [3, 24–26]. The combined research has shown that PMAA undergoes a dramatic conformational change between pH 4 and 6, (corresponding to an α (degree of ionisation) between 0.1 and 0.3), whilst PAA adopts a relatively smooth swelling process in the same pH range (initiating at the same degree of ionisation). In acidic media, due to the increased hydrophobicity, PMAA adopts a globular, contracted structure designed to minimise unfavourable interactions between the hydrophobic backbone and side chain and the aqueous solution, whilst PAA has been described as a random, statistical coil [6, 7, 9]. The PMAA shows significantly increased compaction due to the hydrophobic methyl backbone [8, 13, 22, 24–29], that has been shown to induce *hypercoiling* [8]. This has two net effects—increased hydrophobic density gives it both greater solubilisation potential but at the cost of reduced solubility and mobility.

As the degree of ionisation is increased from pH 4 to 6 the PMAA anionic units begin populating the macromolecule backbone, resulting in a transition between pH 5 and 6 where repulsive units between these charges initiate a macroscopic switch from the compact to the water swollen (described in multiple places as 'rod like' [30, 31]) state. Due to the increased initial compaction in PMAA, this

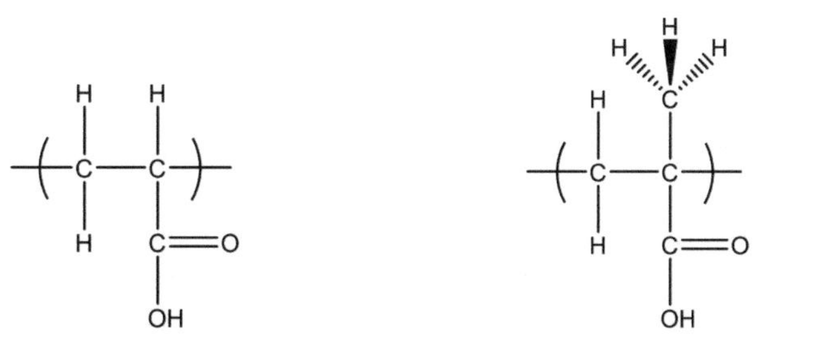

Poly(acrylic acid), PAA   Poly(methacrylic acid), PMAA

Figure 1.
Polyacid chemical structures.

macromolecular swelling results in dramatically changed properties between the compact/swollen polymer. Compared to this, the equivalent deprotonation and subsequent anionic charge drive PAA to adopt an extended state with a relatively smooth transition, with only small changes to polymer physical properties save additional anionic potential. These conformational responses to external stimuli can be viewed as 'smart behaviour' and have led to the incorporation of acrylic acid and methacrylic acid monomers being incorporated into a range of copolymer systems to act as triggers and solvating groups in a range of applications.

Due to the increased compaction, and hydrophobicity, of its globular state, PMAA can solubilise low molar mass organic compounds in solution [12, 17, 32], which is a property not shared by PAA [17, 32, 33]. This is particularly evidenced by the fluorescence emission vibrational fine structure of the aromatic label pyrene. The pyrene excited state emits multiple emission bands, and the relative intensity of bands 1 and 3 vary with different solvents, thus when dispersed in a solution it can give an indication of system polarity [34, 35]. For example, the I_3/I_1 ratio is known to vary between 0.55 (water) and 1.7 (n-pentane) [26]. This feature has been used in the study of many polymer systems, and commonly used by spectroscopists to study macromolecular aggregate structures such as colloids [36], microemulsions [37], micelles [38] and microgels [39, 40]. For example when a 10^{-6} M solution of pyrene was dispersed in an aqueous solution of PAA, the I_3/I_1 ratio did not shift from $\approx$0.55 between pH 3 and 10, identical to the ratio seen for a dispersion in water. This reflects the fact that any interaction between the fluorophore and the polymer does not alter the microenvironment of the label, and confirms the existence of PAA in a water-swollen conformation across the entire pH range. In PMAA at low pH, however, a I_3/I_1 ratio of 1.1 is commonly observed [12], indicating the compact hypercoiled polymer provides hydrophobic shielding from the aqueous solvent. When the pH of pyrene/PMAA solution is increased, this ratio begins to decrease at pH 5, indicating the conformational rearrangement of the polymer, until at pH 6 the probe is released into the solution, returning the fluorescence emission ratio to the state seen in both pure water and PAA. This experiment confirms both the increased solubilisation potential of PMAA over PAA and also the fact that the transition occurs over a broad pH range.

However, the electrostatic potential of these polyelectrolytes cannot be so simply described as indicating that the swollen/collapsed state is neutral/charged as there is an evident near neighbour effect present in polymers that is not seen in comparative small molecule systems [41]. This has been evidenced by the different acid dissociation titration behaviours seen in PMAA when comparing different polymer tacticities [42]. In dilute solutions intrachain interactions across the macromolecule tend to dominate its properties—the molecule can be considered a single long chain surrounded by counter ions, and their solution properties are thus governed by their corresponding electrostatic interactions, which are well described by a range of mathematical theories [43, 44]. To summarise: due to electrostatic repulsion ionisation of acrylate polyelectrolytes occurs over a much wider pH range than observed in the equivalent small molecule, and at the 'stated' pK_a only a fractional ionisation of repeat units will carry a negative charge. For example, potentiometric titrations of PAA found that, at pH 4.5 (pK_a of acrylic acid and the point at which conformational change will occur) only 1/10th of the acrylate repeat units in the polymer will carry this fractional charge [3, 45]. The polymer will continue to ionise up to pH 11 with no further polymer swelling observed despite increasing electrostatic potential of the system. Therefore, it is inappropriate to suggest that the conformational change is driven purely by electrostatic potential, as if this was solely the case further rearrangements at greater degrees of ionisation would be observed (**Figure 2**).

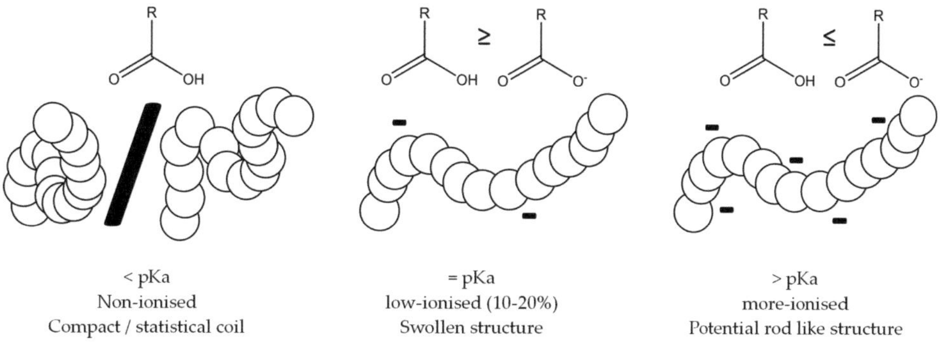

Figure 2.
Conformation of polyacid with different degrees of ionisation.

More recent data indicates that the length scale of the chain plays a role in this transition. For instance, whilst in 0.1 M NaCl the hydrodynamic radii of PAA scales with molar mass [46] the conformational rearrangement of the chai non ionisation in low ionic strength liquids only occurs above a known molar mass lower limit [3]. Current results suggest this is a salt dependent phenomenon [41, 47] and has not been observed in PMAA (although increasing polymer size does slow the kinetics of polymer reconfiguration [27]). As such differences in behaviour between low and high molar mass PAA materials have been observed, such as stark changes in the polymer behaviour at oil-water interfaces [27, 48].

3. Poly(acrylamides)

Not all polyacrylates demonstrate electrolyte properties, and one of the most common non-ionisable acrylate materials produced today is acrylamide copolymers. This chapter concerns itself specifically with two specific materials of particular interest with divergent properties, although there are a range of further examples. These polymers are poly(acrylamide) (PAM) and the hydrophobically modified poly(*N*-isopropylacrylamide) (PNIPAM), whose properties are driven by the additional hydrophobic groups along the polymer side chain. As such one is widely used as an inexpensive, mass-market commodity whilst the other is a very heavily investigated [40, 49], high value material with particular interest in its biomedical applications [50] (**Figure 3**).

Random copolymers of acrylamide (both anionic, cationic and neutrally charged) have been extensively used in the water industry for many years [51–54]. They are extensively employed to remove dissolved organic matter (DOM) for water clarification purposes. Flocculation of fine particles can occur *via* several mechanisms including polymer bridging, charge neutralisation, polymer-particle complex formation and depletion flocculation; often a combination of several of these processes [55]. Binding in poly(acrylamide) is primarily by hydrogen bonding [56], although copolymerised sections may also assist with electrostatic interaction or ion binding. In a sufficiently long polymer chain, there are many potential binding sites, and once sufficient repeat units along a single polymer chain have adhered to a particle surface, the adsorption is often considered irreversible despite the fact each individual binding site is acting in an equilibrium [53]. Once a polymer has adhered to a particle, it can be divided into three segments: *trains* (adhered to the particle surface), *loops* (that extend from the surface) and *tails* (which project into the solution). The speed by which the polymer shifts is difficult to assess but an

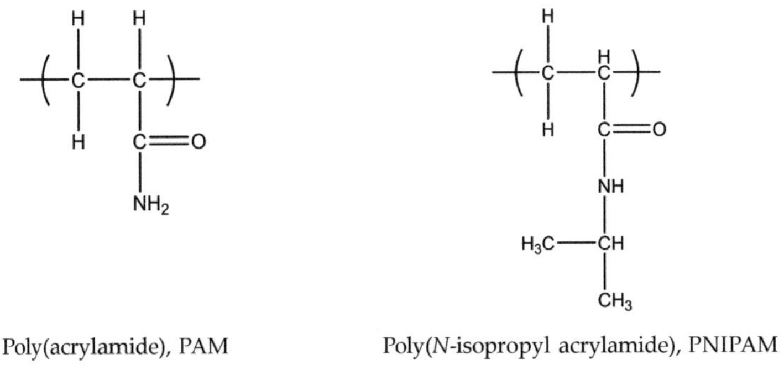

Figure 3.
Polyacylamide chemical structures.

important factor in flocculation kinetics [53]. Following the adhesion of polymer chain of sufficient length for a loop or tail to extend into the solvent, secondary attachment to secondary particles can occur in a process known as 'polymer bridging'. Polymer chains adsorbed on the particle surfaces via only a few points of attachment leave the majority of the chain in solution whilst increasing adsorption onto the particle to saturation reduces the flocculation potential of the polymer. Bridging can be impacted by both the charge density and the molecular weight of the polymer [55] (**Figure 4**).

Polyacrylamide is used in a range of other scenarios including erosion control [51, 57–59], medical implants [60–63], and reduction of water seepage via increasing aqueous viscosity to both stabilise soil and dust prevention [51]. Poly(acrylamide) was one of the first polymers used to reduce soil losses in furrow irrigation [64] and the polymer has been sold commercially for this purpose since at least 1995 [51]. Large quantities of this material are therefore escaping into the environment [59, 65] and a body of research is being built up regarding its effect on the ecosystem [66]. Generally, the polymer is considered non-toxic, with most concerns around its use arising due to its close association from the potent neurotoxin monomer from which it is formed. Since Swedish researchers discovered that acrylamide can be found in heated foodstuffs [67–72], there has been low level public concern about the use of polyacrylamides in a range of industries.

However, studies of polyelectrolyte flocculants of all types have been carried out and consistently poly(acrylamide) is identified as being the primary 'toxicant' [66]. Within much of the poly(acrylamide) literature, there has been a lot of emphasis placed on the toxicity of the monomer, resulting in studies discounting the effects of the polymer and only focusing on residual monomer spread [65]. However, some studies have shown that poly(acrylamide) is unlikely to degrade into residual monomers, or any other toxic compounds [73], and this has only been observed under specific harsh conditions [74]. Therefore, a complete study of the environmental impact of these polymers should include the raw polymeric product.

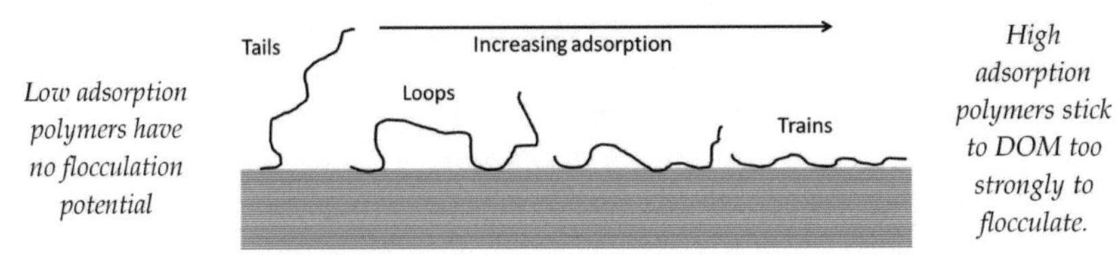

Figure 4.
Increasing polymer adsorption to surfaces.

Testing of poly(acrylamide) interaction with the gill tissues of several aquatic species including fish [66, 73, 75–77], crustaceans [66, 73, 77], algae [66, 77] and insects [77] have been carried out. In many studies of adult fish, the anionic and non-ionic form of the poly(acrylamide) cause only low levels of damage to the fish, with effects increasing at higher concentrations [75]. However, sustained exposure of organisms over a 40 day period has shown that low levels of these polymers are intrinsically toxic to almost all aquatic fauna [73]. Environmental exposure is unlikely to be sustained over long periods due to the polymer desorption to organic matter but few studies have been undertaken into the metabolic rate at which they are removed from living organisms. Even in tests where fish survival was not impacted, the general activity and swimming behaviour of the fish were sub normal [76]. Conversely the cationic form of the polymer is known to be far more toxic, causing pathological issues at sub mg ml^{-1} concentrations, as the polymer builds up on negatively charged gill surfaces [75, 77]. Reduced gill functionality impairs oxygen uptake in the fish and results in death. Further studies have shown that polyelectrolytes can cause adverse changes in fish organ cells (liver and kidneys) [73], decrease animal locomotion and greatly increase respiratory rate. This suggests that the presence of dissolved flocculants may not be lethally toxic but suggests it is capable of causing the fish elevated levels of distress. In invertebrates, their mechanical action was reduced as polyelectrolytes adsorbed onto their body surfaces, reducing their vital functions [73], and again the cationic form of the polymer is far more toxic than the anionic form. [77] In microcosms tests, it has been shown that high polyelectrolyte concentration can reduce algal growth [66, 77]. This in turn can increase the potential toxicity of the polymer as the algae acts as a neutralising agent towards the polymer. To algae, even the anionic and non-ionic polymer is toxic, negatively affecting both cell growth and O_2 production [73]. It has been observed that addition of combinations of both anionic and cationic polymer can reduce toxicity [77] and several patents have been issued suggesting that anionic polymers can be used to detoxify cationic polyelectrolytes [78, 79]. In conclusion, the discrepancy between anionic and cationic polymers in regard to aquatic toxicity must be considered in the application of these polymers [75, 77]. The cationic form of the polymer is regarded as generally more toxic but the anionic form has also been shown to cause chronic, sub-lethal responses even at low concentrations [66].

Although it has some larger applications, PNIPAM is not produced or utilised in such great quantities. It is mainly of interest due to a thermally induced conformation the polymer exhibits at 32°C, caused by the hydrophobic isopropyl groups [49]. This ‘smart’ response has led to great interest in the polymer, both to understand its properties and apply them in a range of fields, specifically in Bioengineering [50]. In essence, the polymer has a lower critical solution temperature (LCST), a conformational change that occurs *via* a two stage process. Firstly, the polymer has an intramolecular collapse, where individual chains contract in upon themselves as they break hydrogen bonds with the aqueous solution, followed by a secondary event of intramolecular aggregation of the collapsed coils [80]. This event is triggered by the increasing entropic cost at high temperature of the restricted water that solubilises the dissolved polymer chains below the LCST. The event has some hysteresis between heating and cooling radii of gyration [81], governed by two intermediate states that give PNIPAM four potential conformations: globule, molten globule crumpled coil and coil [82]. During the collapse the globular state dispenses approximately 34% of the water molecules [82], meaning that although this is a desolvation event leading to an insoluble material, collapsed PNIPAM can never be described as a hydrophobic system (**Figure 5**) [49].

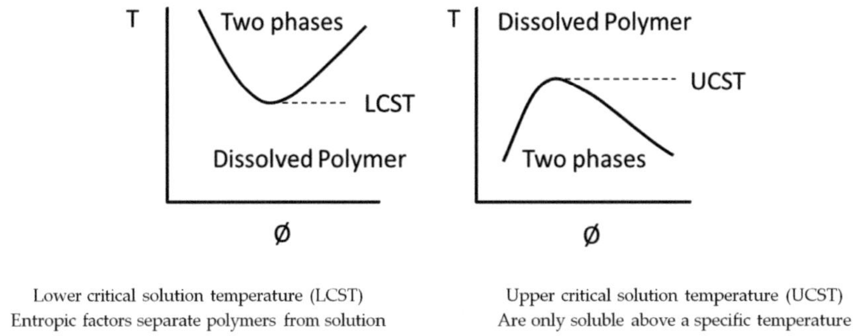

Figure 5.
Typical thermoresponsive properties of polymers [83]: (temperature (T) vs. volume fraction (Ø)).

The LCST of PNIPAM can be affected by the addition of hydrophobic or hydrophilic end groups [84], or the molecular weight and concentration of the sample [85]. Due to the LCST is reasonably close to body temperature, there has been much work to manipulate PNIPAM to act as a drug-delivery agent or trigger or apply it in other bio-engineering circumstances [86, 87].

4. Acrylate interpolymer complexes

Interactions between multiple polymers in a formulation are almost inevitable, and there has been plenty of studies of specific driving factors undertaken over the last 50 years to build a strong picture of inter-polymer interactions. This phase separation phenomena is observed in even the most dilute solutions, as it is driven by a mixture of electrostatic, hydrogen bonding and hydrophobic interactions, all dependent on pH, salt concentration and temperature [88, 89]. To our knowledge this type of complex was first patented in 1966 [90], with much of the following fundamental measurements carried out over the following decades [14, 91–94]. Since then, the system has been described as a *laddered* sequences of bonds between the molecules, occasionally interrupted with *loop* defects [95], an evolution similar to the model previously described of polymer adhesion to surfaces. This theory originally posed that the polymers will form rigid, static structures due to repeated hydrogen-bonding across molecules. More recent studies have put less emphasis on the polymer rigidity and have given an alternative description of these repeated labile interactions more as *ribbons* (i.e., two flexible materials that can slide over each other).

The interactions between PAA and PAM are one of the more studied systems of interpolymer complex formation (IPC) [93, 96–101], and in both solution and solid state the interaction has been shown to be pH dependant [14, 96]. Mixed solutions of PAA and PAM form a turbid solution that precipitates when cooled [97]. This phase separation follows the formation of complexes between PAA and PAM that varies in structure depending on the concentration, medium and the ionisation constant [97]. For complexes between PAA and a proton-acceptor polymer it has been shown that IPCs will only form below a critical value of pH (pH_{crit}) [14, 99], the structure dependent point above which any partial neutralisation of the polyacid inhibits complex formation [14, 100].

Early work within this field required high molecular weight materials to detect complex formation [99, 102], however, modern instrumentation has facilitated detection of smaller complexes down at the parts per million loading level [103]. The structure of the resultant IPC (whether in a gel or a compact solvated complex) depends on the relative molecular weight of complexing partners, [104] but as this is a multivalent effect of repeated binding sites, larger molecular weight materials demonstrate stronger interactions. Furthermore, it has been indicated that very large molecular weight polyacids have been seen to raise pH_{crit} [93].

When dissolved in high ionisation solutions, both polymers have rapid segmental motion, existing as random polymeric coils. If the solution ionisation is decreased, this deprotonates the acidic polyelectrolytes and reduces its affinity for inter-polymer complexation. This occurs as can now form both intramolecular H-bonds internally across the chain backbone or intermolecularly forming H-bonds with other polymers [105], leading to a rigid polymer mixture with restricted chain motions. PAA forms stronger complexes to PAM than some other polymers (i.e., poly(ethylene oxide) or poly(vinyl acetate)) due to additional ion-dipole interaction of the partially protonated amide groups and the C=O dipoles of PAA [106]. In ambient conditions the peak aqueous interaction between PAA and PAM occurs ≈ pH 2.69 [107] but this is affected by many environmental factors including temperature [89, 101], ionic strength [93, 107, 108] and the addition of inorganic binders [99]. The complex polymer/polymer/solvent ratio of interactions is temperature sensitive causing PAM-PAA copolymers to become upper-critical solution temperature materials (an inversion of the LCST seen in PNIPAM where they become only soluble above a specific temperature [87]). These combined external factors deteriorate the thermodynamic quality of the solvent, strengthening polymer-polymer interactions by weakening polymer-water solvation [109]. However, below pH_{crit}, only small portions of the PAM form into 'multimacroion clusters', indicating that in an equivalent system with 1:1 acid/acrylamide repeat units, a large percentage of acrylamide will be free in solution unbound to PAA [110]. This was the first result of several which have cast doubt on the ladder model, and computational modelling software of polymer/polymer ionic interactions has proposed a range of complexing structures ranging from ladders to scrambled egg structures [111]. Further experimental evidence has shown that a PAA coil does not unwind or swell on addition of a PAM polymer but potentially contracts into a smaller co-globule [112], and an explanation for this can be found when considering the difference between the pH_{crit} of IPC formation and pK_a of PAA conformational change.

Other acrylate materials, such as PNIPAM, demonstrate similar responses to polyacids, and exhibit their own IPC potential [113]. Whilst PAM-PAA interactions are dominant at lower temperatures, PAA-PNIPAM show increased interactions at high temperatures, indicating that the complex formation is driven by hydrophobic interactions not seen in the base acrylamide structure [114]. Studies using dissolved pyrene indicated these lead to stronger interactions between PAA and PNIPAM than PAA-PAM [115, 116]. Furthermore, the hydrophobic isopropyl side chain causes PNIPAM to alter its response to ionic strength. Whereas PAM-PAA complexes are strengthened by increasing ionic strength, PNIPAM complexes show decreased critical pH reducing their bond forming potential [109]. As the initial critical pH for IPC formation was larger than 3, Khutoryanskiy theorised that the increasing ionic strength partially dissociates the polyacid. As only non-ionised carboxylic groups are able to form hydrogen bonds, this impedes IPC formation and reduces the pK_a.

For PAA-PAM, pH_{crit} was found to be 2.7, whilst for increasing hydrophobic additions to the acrylamide unit (poly(ethyl acrylamide), poly(dimethyl acrylamide), poly(diethyl acrylamide), the pH_{crit} was found to increase to 3, 4, 5 respectively [14]. It is interesting to note that the most hydrophilic acrylamide polymers (including a hydrophilic-functionalised poly(hydroxyl ethyl acrylamide)) show lower pH_{crit}, indicating that their IPC with the polyacid are less tolerant of deprotonation. The PAA-PAM complex appears to separate when the first acid repeat unit

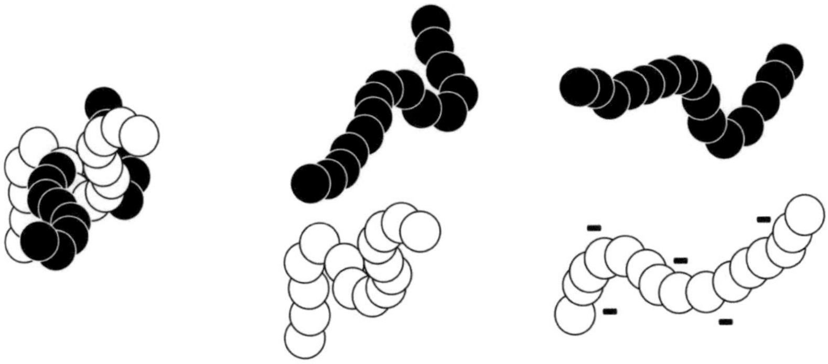

Figure 6.
Three conformational states of PAA-PAM in mixed solutions.

along the chain is deprotonated whilst more hydrophobically modified polymers are more tolerant to partial ionisation when complexing with PAA. Computational modelling of the solvation energy of each repeat unit shows a clear correlation between solvation potential and pH_{crit} [14].

Clearly the polyacid dictates the potential of IPC with receptive polymers [96], and in this mind, it is worth revisiting the PAA-PAM IPC structure. As the pK_a of the PAA is higher than pH_{crit}, it will be 'non-swollen' before it encounters a complexing partner. Early literature in the subject, comparing the supposed 'non-response' of PAA compared to PMAA cited, its only-slight alteration to solution viscosity and inability to solubilise hydrophobic dyes as evidence it had no conformational response. However, more recent studies with more sensitive techniques have shown that this is not the case and PAA does indeed go through a lesser swelling-contraction event. As such, it is proposed that the PAA can exist in three potential conformations in the presence of a polymeric bonding partner (**Figure 6**).

We suggest that compacted PAA has no entropic or enthalpic reason to uncoil or swell prior to complexation. Given the combined evidence from two separate sources that (1) most PAM chains are not binding to partners and (2) PAA chains do not swell further apart on PAM binding (in fact there is some evidence of contraction), it seems reasonable to propose that the PAA-PAM complex is not amorphous in nature, and certainly not an extended ribbon/ladder structure.

5. Conclusion

This chapter reviews some of the recent developments in polyacrylate properties and interactions, and delves deeply into their industrial applications to provide both further context and understanding. During the early study many assumptions were made due to the difficulty to analyse these large macromolecules, particularly in dilute solutions, and our understanding of these systems has slowly evolved as more advanced technology with greater sensitivity has facilitated deeper interrogation of these systems [117]. This chapter only touches on a few choice themes of polymer-responsiveness and ignored many of the more challenging aspects of the field. The state of ionisation of all of these polymers has clearly been shown to have an effect on their solution properties, and although the field is still under development after several decades of work, common themes can be seen across the subject dictating macromolecular conformational changes.

Acknowledgements

This book chapter is both an update, summation of, and substantial revision to my Ph.D., work area carried out at the University of Sheffield, originally under Dr. Linda Swanson, who has published extensively on the photophysical analytical techniques described in this paper. For further reference on the application of those techniques in these systems, particularly in the study of polyelectrolytes, please see her book chapter 'Optical Properties of Polyelectrolytes' [26].

Conflict of interest

There are no conflicts of interest to declare.

Thanks

Written following the arrival of, and dedicated to, Jonathan Swift, born in 2017.

Abbreviations

IPC	interpolymer complex formation
DNA	deoxyribonucleic acid
DOM	dissolved organic matter
PAA	poly(acrylic acid)
PAM	poly(acrylamide)
PMAA	poly(methacrylic acid)
PNIPAM	poly(*N*-isopropylacrylamide)

Author details

Thomas Swift
University of Bradford, Bradford, United Kingdom

*Address all correspondence to: t.swift@bradford.ac.uk

References

[1] Polyacrylate Market - Segmented by Product Type, Application, and Geography - Trends and Forecast (2018-2023). Hyderabad: Mordor Intelligence; 2018

[2] Harper BG, Niles Bashaw R, Leroy Atkins B. Absorbent Product Containing a Hydrocellodal Composition. Patent US3670731; 1966

[3] Swift T, Swanson L, Geoghegan M, Rimmer S. The pH-responsive behaviour of poly(acrylic acid) in aqueous solution is dependent on molar mass. Soft Matter. 2016;**12**:2542-2549

[4] Katchalsky A, Spitnik P. Potentiometric titrations of polymethacrylic acid. Journal of Polymer Science. 1947;**2**:432-446

[5] Leyte JC, Mandel M. Potentiometric behavior of polymethacrylic acid. Journal of Polymer Science, Part A: General Papers. 1964;**2**:1879-1891

[6] Crescenzi V. Some recent studies of polyelectrolyte solutions. Advances in Polymer Science. 1968;**5**:358-386

[7] Arnold R. The titration of polymeric acids. Journal of Colloid Science. 1957;**12**:549-556

[8] Katchalsky A. Solutions of polyelectrolytes and mechanochemical systems. Journal of Polymer Science, Part A: General Papers. 1951;**7**:393-412

[9] Mandel M. The potentiometric titration of weak polyacids. European Polymer Journal. 1970;**6**:807-822

[10] Přádný M, Holata J, Ševčík S. The use of low-molecular-weight model compounds in an investigation of polyelectrolytes, 1. Potentiometric properties of poly(methacrylic acid) and its 2-dimethylaminoethyl ester. Macromolecular Chemistry and Physics. 1989;**190**:1079-1088

[11] Eisenberg H. Conductance of partially neutralized polymethacrylic and polyacrylic acids, using a polarization compensated twin cell. Journal of Polymer Science, Part A: General Papers. 1958

[12] Chen TS, Thomas JK. Influence of the conformational state of polymethacrylic acid on the photophysical properties of pyrene in aqueous solution: A fluorescent probe and laser photolysis study. Journal of Polymer Science, Part A: Polymer Chemistry. 1979;**17**:1103-1116

[13] Soutar I, Swanson L. Luminescence studies of polyelectrolyte behaviour in solution-I. Accessibility of naphthalene-based labels of poly(methacrylic acid) to mobile low molar mass species in aqueous media. European Polymer Journal. 1993;**29**:371-378

[14] Swift T, Seaton CC, Rimmer S. Poly(acrylic acid) interpolymer complexes. Soft Matter. 2017;**14**:8736-8744

[15] Alves L, Lindman B, Klotz B, Böttcher A, Haake HM, Antunes FE. Rheology of polyacrylate systems depends strongly on architecture. Colloid and Polymer Science. 2015;**239**:3285-3293

[16] Shallcross L, Roche K, Wilcock CJ, Stanton KT, Swift T, Rimmer S, et al. The effect of hyperbranched poly(acrylic acid)s on the morphology and size of precipitated nanoscale (fluor)hydroxyapatite. Journal of Materials Chemistry B. 2017;**5**:6027-6033

[17] Barone G, Crescenzi V, Liquori AM, Quadrifoglio F. Solubilization of polycyclic aromatic hydrocarbons in poly(methacrylic acid) aqueous solutions. The Journal of Physical Chemistry. 1967;**71**:2341-2345

[18] Morawetz H. Macromolecules in Solution. Wiley; 1975

[19] Nekrasova TN, Anufriyeva YV, Yel'yashevich AM, Ptitsyn OB. Potentiometric titration of polyacrylic acid, polymethacrylic acid and poly-L-glutamic acid. Polymer Science USSR. 1965;7:1008-1018

[20] Koenig JL, Angood AC, Semen J, Lando JB. Laser-excited Raman studies of the conformational transition of syndiotactic polymethacrylic acid in water. Journal of the American Chemical Society. 1969;**91**:7250-7254

[21] Heitz C, Rawiso M, François J. X-ray scattering study of a poly(methacrylic acid) sample as a function of its neutralization degree. Polymer (Guildf). 1999;**40**:1637-1650

[22] Pleštil J, Ostanevich YM, Bezzabotonov VY, Hlavatá D, Labský J. Small-angle scattering from polyelectrolyte solutions: Dimensions of poly(methacrylic acid) chains in salt-free solutions. Polymer (Guildf). 1986;**27**:839-842

[23] Moussaid A, Schosseler F, Munch JP, Candau SJ. Structure of polyacrylic acid and polymethacrylic acid solutions: A small angle neutron scattering study. Journal of Physics B Atomic and Molecular Physics. 1993;**3**:573-594

[24] Ruiz-Pérez L, Pryke A, Sommer M, Battaglia G, Soutar I, Swanson L, et al. Conformation of poly(methacrylic acid) chains in dilute aqueous solution. Macromolecules. 2008;**41**:2203-2211

[25] Olea AF, Thomas JK. Fluorescence studies of the conformational changes of poly(methacrylic acid) with pH. Macromolecules. 1989;**22**:1165-1169

[26] Swanson L. Optical properties of polyelectrolytes. In: Photochemistry and Photophysics of Polymer Materials. Hoboken, New Jersey: John Wiley & Sons inc.; 2010. pp. 41-92

[27] Olea AF, Rosenbluth H, Thomas JK. Effect of the molecular weight on the dynamics of the conformational transition of poly(methacrylic acid). Macromolecules. 1999;**27**:8077-8083

[28] Chu DY, Thomas JK. Photophysical studies of a water-soluble copolymer of methacrylic acid and 1-pyreneacrylic acid. Macromolecules. 1984;**17**:2142-2147

[29] Bednar B, Morawetz H, Shafer JA. Kinetics of the conformational transition of poly(methacrylic acid) after changes of its degree of ionization. Macromolecules. 1985;**18**:1940-1944

[30] Katchalksy A, Eisenberg H. Molecular weight of polyacrylic and polymethacrylic acid. Journal of Polymer Science. 1951;**6**:145-154

[31] Soutar I, Swanson L. Luminescence studies of polyelectrolyte behavior in solution. 3. Time-resolved fluorescence anisotropy measurements of the conformational behavior of poly(methacrylic acid) in dilute aqueous solutions. Macromolecules. 1994;**27**:4304-4311

[32] Tan KL, Treloar FE. Solubilization of 9-methylanthracene by the hypercoiled form of poly(methacrylic acid) in water: Fluorescence decay and rotational diffusion measurements. Chemical Physics Letters. 1980;**73**:234-239

[33] Treloar FE. Conformational transition in poly(methacrylic acid) in aqeous-solution-dye binding and fluorescence depolarization. Chemica Scripta. 1976;**10**:219-224

[34] Akira N, Hiroaki B. Fluorescence spectrum of pyrene vapor: Emission from the second excited singlet state. Bulletin of the Chemical Society of Japan. 1970;**43**:967-967

[35] Kalyanasundaram K, Thomas JK. Environmental effects on vibronic

band intensities in pyrene monomer fluorescence and their application in studies of micellar systems. Journal of the American Chemical Society. 1977;**99**:2039-2044

[36] Soutar I, Swanson L, Annable T, Padget JC, Satgurunathan R. Luminescence techniques and characterization of the morphology of polymer latices. 3. An investigation of the microenvironments within stabilized aqueous latex dispersions of poly(n-butyl methacrylate) and polyurethane. Langmuir. 2006;**22**:5904-5910

[37] Lianos P, Lang J, Zana R. Fluorescence probe study of oil-in-water microemulsions. 2. Effect of the nature of alcohol, oil, and surfactant on the surfactant aggregation number in the aggregates. The Journal of Physical Chemistry. 1982;**86**:4809-4814

[38] Jay J, Johnston L, Scaiano JC. Quenching of pyrene fluorescence by cupric ions in micellar solution: Effect of quenching on the polarity reported by the prob. Chemical Physics Letters. 1988;**148**:517-522

[39] Pankasem S, Thomas JK, Snowden MJ, Vincent B, Snowden MJ. Photophysical studies of poly (*N*-isopropylacrylamide) microgel structures. Langmuir. 1994;**10**:3023-3026

[40] Flint NJ, Gardebrecht S, Swanson L. Fluorescence investigations of "smart" microgel systems. Journal of Fluorescence. 1998;8:343-353

[41] Dolce C, Mériguet G. Ionization of short weak polyelectrolytes: When size matters. Colloid & Polymer Science. 2017

[42] Loebl EM, O'Neill JJ. Solution properties of isotactic polymethacrylic acid. Journal of Polymer Science, Part A: General Papers. 1960;**45**:538-540

[43] Dobrynin AV, Rubinstein M. Theory of polyelectrolytes in solutions and at surfaces. Progress in Polymer Science. 2005;**30**:1049-1118

[44] Kuhn W, Künzle O, Katchalsky A. Verhalten polyvalenter Fadenmolekelionen in Lösung. Helvetica Chimica Acta. 1948;**31**:1049-118

[45] Anghel DF, Alderson V, Winnik FM, Mizusaki M, Morishima Y. Fluorescent dyes as model "hydrophobic modifiers" of polyelectrolytes: A study of poly(acrylic acid)s labelled with pyrenyl and naphthyl groups. Polymer (Guildf). 1998;**39**:3035-3044

[46] Reith D, Müller B, Müller-Plathe F, Wiegand S. How does the chain extension of poly (acrylic acid) scale in aqueous solution? A combined study with light scattering and computer simulation. The Journal of Chemical Physics. 2002;**116**:9100

[47] Geoghegan M. The swelling of weak polyelectrolytes at low salt concentrations in dilute solution. Polymer (United Kingdom). 2017;**112**:414-417

[48] Zaibudeen AW, Philip J. A spectroscopic approach to probe macromolecular conformational changes at interface under different environmental conditions: A case study with PAA adsorbed at oil-water interface. Journal of Molecular Liquids. 2018;**252**:30-39

[49] Pelton R. Poly(*N*-isopropyl-acrylamide) (PNIPAM) is never hydrophobic. Journal of Colloid and Interface Science. 2010;**348**:673-674

[50] Shepherd J, Sarker P, Rimmer S, Swanson L, MacNeil S, Douglas I. Hyperbranched poly(NIPAM) polymers modified with antibiotics for the reduction of bacterial burden in infected human tissue engineered skin. Biomaterials. 2011;**32**:258-267

[51] Sojka RE, Bjorneberg DL, Entry JA, Lentz RD, Orts WJ. Polyacrylamide in agriculture and environmental land management. Advances in Agronomy. 2007;**92**:75-163

[52] Yongrui P, Zheng Z, Bao M, Li Y, Zhou Y, Sang G. Treatment of partially hydrolyzed polyacrylamide wastewater by combined Fenton oxidation and anaerobic biological processes. Chemical Engineering Journal. 2015;**273**:1-6

[53] Bolto B, Gregory J. Organic polyelectrolytes in water treatment. Water Research. 2007;**41**:2301-2324

[54] Aguilar MI, Sáez J, Lloréns M, Soler A, Ortuño JF. Nutrient removal and sludge production in the coagulation-flocculation process. Water Research. 2002;**36**:2910-2919

[55] Nasser MS, James AE. The effect of polyacrylamide charge density and molecular weight on the flocculation and sedimentation behaviour of kaolinite suspensions. Separation and Purification Technology. 2006;**52**:241-252

[56] Griot O, Kitchener JA. Role of surface silanol groups in the flocculation of silica suspensions by polyacrylamide: Part 2—Surface changes of silica suspensions on ageing. Transactions of the Faraday Society. 1965;**61**:1026-1031

[57] Kang J, Sowers TD, Duckworth OW, Amoozegar A, Heitman JL, McLaughlin RA. Turbidimetric determination of anionic polyacrylamide in low carbon soil extracts. Journal of Environmental Quality. 2013;**42**:1902-1907

[58] Trout TJ, Sojka RE, Lentz RD. Polyacrylamide effect on furrow erosion and infiltration. American Society of Agricultural and Engineers. 1995;**38**:761-765

[59] Lentz RD, Sojka RE, Foerster JA. Estimating polyacrylamide concentration in irrigation water. Journal of Environmental Quality. 1996;**25**:1015-1024

[60] Christensen LH, Breiting VB, Aasted A, Jørgensen A, Kebuladze I. Long-term effects of polyacrylamide hydrogel on human breast tissue. Plastic and Reconstructive Surgery. 2003;**111**:1883-1890

[61] Cheng NX, Liu LG, Hui L, Chen YL, Xu SL. Breast cancer following augmentation mammaplasty with polyacrylamide hydrogel (PAAG) injection. Aesthetic Plastic Surgery. 2009;**33**:563

[62] Davis BK. Control of diabetes with polyacrylamide implants containing insulin. Experientia. 1972;**28**:348

[63] Altman D, Hjern F, Zetterström J. Transanal submucosal polyacrylamide gel injection treatment of anal incontinence: A randomized controlled trial. Acta Obstetricia et Gynecologica Scandinavica. 2016;**95**:528-533

[64] Lentz RD, Sojka RE, Carter DL, Shainberg I. Preventing irrigation furrow erosion with small applications of polymers. Soil Science Society of America Journal. 1992;**56**:1926-1932

[65] Touzé S, Guerin V, Guezennec AG, Binet S, Togola A. Dissemination of acrylamide monomer from polyacrylamide-based flocculant use—Sand and gravel quarry case study. Environmental Science and Pollution Research. 2015;**22**:6423-6430

[66] Harford AJ, Hogan AC, Jones DR, van Dam RA. Ecotoxicological assessment of a polyelectrolyte flocculant. Water Research. 2011;**45**:6393-6402

[67] Yaylayan VA, Wnorowski A, Perez Locas C. Why asparagine needs carbohydrates to generate acrylamide.

Journal of Agricultural and Food Chemistry. 2003;**51**:1753-1757

[68] Tareke E, Rydberg P, Karlsson P, Eriksson S, Törnqvist M. Acrylamide: A cooking carcinogen? Chemical Research in Toxicology. 2000;**13**:517-522

[69] Becalski A, Lau BPY, Lewis D, Seaman SW. Acrylamide in foods: Occurence, sources and modelling. Journal of Agricultural and Food Chemistry. 2003;**51**(3):802-808

[70] Zyzak DV, Sanders RA, Stojanovic M, Tallmadge DH, Eberhart BL, Ewald DK, et al. Acrylamide formation mechanism in heated foods. Journal of Agricultural and Food Chemistry. 2003;**51**:4782-4787

[71] Ahn JS, Castle L, Clarke DB, Lloyd AS, Philo MR, Speck DR. Verification of the findings of acrylamide in heated foods. Food Additives and Contaminants. 2002;**19**:1116-1124

[72] Rydberg P, Eriksson S, Tareke E, Karlsson P, Ehrenberg L, Törnqvist M, et al. Investigations of factors that influence the acrylamide content of heated foodstuffs. Journal of Agricultural and Food Chemistry. 2003;**51**:7012-7018

[73] Beim AA, Beim AM. Comparative ecological—toxicological data on determination of maximum permissible concentrations (mpc) for several flocculants. United Kingdom: Environmental Technology; 1994;**15**:195-198

[74] Aksberg R, Wågberg L. Hydrolysis of cationic polyacrylamides. Journal of Applied Polymer Science. 1989;**38**:297-304

[75] Kerr JL, Lumsden JS, Russell SK, Jasinska EJ, Goss GG. Effects of anionic polyacrylamide products on gill histopathology in juvenile rainbow trout (*Oncorhynchus mykiss*). Environmental Toxicology and Chemistry. 2014;**33**:1552-1562

[76] Liber K, Weber L, Lévesque C. Sublethal toxicity of two wastewater treatment polymers to lake trout fry (*Salvelinus namaycush*). Chemosphere. 2005;**61**:1123-1133

[77] Biesinger KE, Stokes GN. Effects of synthetic polyelectrolytes on selected aquatic organisms. Journal Water Pollution Control Federation. 1986;**58**:207-213

[78] Furuno Method of rendering cationic polymers harmless to fish. 1976

[79] Furuno. Process for the Detoxification of Water Treated with Polymer Cationic Flocculation Agents. 1980

[80] Chee CK, Rimmer S, Soutar I, Swanson L. Fluorescence investigations of the thermally induced conformational transition of poly(*N*-isopropylacrylamide). Polymer (Guildf). 2001;**42**:5079-5087

[81] Wang X, Qiu X, Wu C. Comparison of the coil-to-globule and the globule-to-coil transitions of a single poly (*N*-isopropylacrylamide) homopolymer chain in water. Macromolecules. 1998;**31**:2972-2976

[82] Schmaljohann D. Thermo- and pH-responsive polymers in drug delivery. Advanced Drug Delivery Reviews. 2006;**58**:1655-1670

[83] Kammer H-W, Inoue T, Ougizawa T. Upper and lower critical solution temperature behaviour in polymer blends and its thermodynamic interpretation. Polymer (Guildf). 1989;**30**:888-892

[84] Kujawa P, Aseyev V, Tenhu H, Winnik FM. Temperature-sensitive properties of poly(*N*-isopropylacrylamide) mesoglobules formed in dilute aqueous solutions heated above their demixing point. Macromolecules. 2006;**39**:7686-7693

[85] Plunkett KN, Zhu X, Moore JS, Leckband DE. PNIPAM chain collapse depends on the molecular weight and grafting density. Langmuir. 2006;**22**:4259-4266

[86] Ward MA, Georgiou TK. Thermoresponsive polymers for biomedical applications. Polymers (Basel). 2011;**3**:1215-1242

[87] Gandhi A, Paul A, Sen SO, Sen KK. Studies on thermoresponsive polymers: Phase behaviour, drug delivery and biomedical applications. Asian Journal of Pharmaceutical Sciences. 2015;**10**:99-107

[88] Bailey FE, Lukdberg RD, Callard RW. Some factors meeting the molecular association of poly(ethylene oxide) and poly (acrylic acid) In aqueous solution. Journal Polymer Science PART A. 1964;**2**:845-851

[89] Sudre G, Tran Y, Creton C, Hourdet D. PH/Temperature control of interpolymer complexation between poly(acrylic acid) and weak polybases in aqueous solutions. Polymer (Guildf). 2012;**53**:379-385

[90] Smith KL, Winslow AE, Seltzer EC. Chemical reaction product of polycarboxylic acid and a polymeric polyether. Patent US3387061A; 1966

[91] Ikawa T, Abe K, Honda K, Tsuchida E. Interpolymer complex between poly(ethylene oxide) and poly(carboxylic acid). Journal of Polymer Science, Polymer Chemistry Edition. 1975;**13**:1505-1514

[92] Tsuchida E, Osada Y, Ohno H. Formation of interpolymer complexes. Journal Macromolecular Science Part B. 1980;**17**:683-714

[93] Sivadasan K, Somasundaran P, Turro NJ. Fluorescence and viscometry study of complexation of poly(acrylic acid) with poly(acrylamide) and hydrolysed poly(acrylamide). Colloid & Polymer Science. 199;**269**:131-137

[94] Abe K, Koide M, Tsuchida E. Selective complexation of macromolecules. Macromolecules. 1977;**10**:1259-1264

[95] Baranovsky VY, Kazarin LA, Litmanovich AA, Papisov IM. Thermochemical reactions in polycomplexes. European Polymer Journal. 1984;**20**:191-194

[96] Garces FO, Sivadasan K, Somasundaran P, Turro NJ. Interpolymer complexation of poly(acrylic acid) and polyacrylamide: Structural and dynamic studies by solution- and solid-state NMR. Macromolecules. 1994;**27**:272-278

[97] Klenina OV, Fain EG. Phase separation in the system polyacrylic acid-polycrylamide-water. Polymer Science USSR. 1981;**23**:1439-1446

[98] Swift T, Swanson L, Rimmer S. Poly(acrylic acid) interpolymer complexation: Use of a fluorescence time resolved anisotropy as a poly(acrylamide) probe. RSC Advances. 2014;**4**:57991-57995

[99] Mun GA, Nurkeeva ZS, Khutoryanskiy VV, Sarybayeva GS, Dubolazov AV. pH-effects in the complex formation of polymers I. Interaction of poly(acrylic acid) with poly(acrylamide). European Polymer Journal. 2003;**39**:1687-1697

[100] Staikos G, Bokias G, Tsitsilianis C. The viscometric methods in the investigation of the polyacid-polybase interpolymer complexes. Journal of Applied Polymer Science. 1993;**48**:215-217

[101] Staikos G, Karayanni K, Mylonas Y. Complexation of polyacrylamide and poly(*N*-isopropylacrylamide) with poly(acrylic acid). The temperature

effect. Macromolecular Chemistry Physics. 1997;**198**:2905-2915

[102] Eustace DJ, Siano DB, Drake EN. Polymer compatibility and interpolymer association in the poly(acrylic acid)-polyacrylamide-water ternary system. Journal of Applied Polymer Science. 1988;**35**:707-716

[103] Swift T, Swanson L, Bretherick A, Rimmer S. Measuring poly(acrylamide) flocculants in fresh water using inter-polymer complex formation. Environmental Science: Water Research & Technology. 2015;**1**:332-340

[104] Staikos G, Tsitsilianis C. Viscometric investigation of the poly(acrylic acid)—Polyacrylamide interpolymer association. Journal of Applied Polymer Science. 1991;**42**:867-872

[105] Dong J, Ozaki Y, Nakashima K. Infrared, Raman, and near-infrared spectroscopic evidence for the coexistence of various hydrogen-bond forms in poly(acrylic acid). Macromolecules. 1997;**30**:1111-1117

[106] Tsuchida E, Abe K. Interactions between macromolecules in solution and intermacromolecular complexes. In: Interactions between Macromolecules in Solution and Intermacromolecular Complexes. Berlin, Heidelberg: Springer; 1982. p. 88

[107] Moharram MA, Balloomal LS. Infrared study of the complexation of poly(acrylic acid) with poly(acrylamide). Journal of Applied Polymer Science. 1996;**59**:987-990

[108] Nurkeeva ZS, Mun GA, Khutoryanskiy VV, Sergaziev AD. Complex formation between poly(vinyl ether of diethyleneglycol) and polyacrylic acid I. Effect of low molecular salts and phenols additives. European Polymer Journal. 2001;**37**:1233-1237

[109] Khutoryanskiy VV, Mun GA, Nurkeeva ZS, Dubolazov AV. pH and salt effects on interpolymer complexation via hydrogen bonding in aqueous solutions. Polymer International. 2004;**53**:1382-1387

[110] Deng L, Wang C, Li ZC, Liang D. Re-examination of the "zipper effect" in hydrogen-bonding complexes. Macromolecules. 2010;**43**:3004-3010

[111] Lazutin AA, Semenov AN, Vasilevskaya VV. Polyelectrolyte complexes consisting of macromolecules with varied stiffness: Computer simulation. Macromolecular Theory and Simulations. 2012;**21**:328-339

[112] Swift T, Paul N, Swanson L, Katsikogianni M, Förster RS. Resonance energy transfer across interpolymer complexes of poly(acrylic acid) and poly(acrylamide). Polymer (United Kingdom). 2017;**123**:10-20

[113] Garay MT, Alava C, Rodriguez M. Study of polymer-polymer complexes and blends of poly(*N*-isopropylacrylamide) with poly(carboxylic acid). 2. Poly(acrylic acid) and poly(methacrylic acid) partially neutralized. Polymer (Guildf). 2000;**38**:5091-5096

[114] Khutoryanskiy VV, Nurkeeva ZS, Mun GA, Dubolazov AV. Effect of temperature on aggregation/dissociation behavior of interpolymer complexes stabilized by hydrogen bonds. Journal of Applied Polymer Science. 2004;**93**:1946-1950

[115] Staikos G. Interpolymer complexes of poly(acrylamide) and poly (*N*-isopropylacrylamide) with poly(acrylic acid): A comparative study. Polymer International. 1996;**41**:345-350

[116] Koussathana M, Lianos P, Staikos G. Investigation of hydrophobic interactions in dilute aqueous solutions of hydrogen-bonding interpolymer complexes by steady-state

and time-resolved fluorescence measurements. Macromolecules. 1997;**30**:7798-7802

[117] Khutoryanskiy VV, Smyslov RY, Yakimansky AV. Modern methods for studying polymer complexes in aqueous and organic solutions. Polymer Science, Series A. 2018;**60**(5):553-576

Chapter 2

Parametric Studies on Transmission Laser Welding of Acrylics

Ramesh Rudrapati

Abstract

Due to limitations of injection molding of polymer/plastic materials, complex plastic parts are often assembled from two or more injection-molded components. Joining of polymers can be achieved by chemical-based technologies and thermal methods such as welding. Welding technique is one of the important manufacturing routes that can be used to refine product design and reduce production cost. Plastic welding processes typically involve heating the joining faces to induce localized melting and subsequently applying pressure to cause molecular diffusion at the molten interface, which produces a solid weld upon cooling. The need for an effective welding process that is fast, accurate, and with no relative part movement has fueled the development of laser-transmission welding (LTW) technology. LTW is an innovative joining process for acrylate materials. The quality of weld highly depends on correct selection of process parameters in LTW. The systematic study and analysis are required to conduct LTE process economically and efficiently. In the present chapter, current prospects of applications of acrylates and joining of them using LTW has been analyzed. The main emphasis has been given to analyze the variations of quality performance characteristics with varying input welding factors and concluding remarks has been drawn from present work. From this study, it is observed that acrylics are future innovative industrial materials, which need to be joined to create complex features on them. Welding of acrylics using LTW to achieve better and more economical weld performance is still under continuous research by scientists/industrialists.

Keywords: acrylics, plastic welding, laser transmission welding, laser welding parameters

1. Introduction

Manufacturing is a methodology that is used to transform the raw materials into useful products by performing series of operations on it [1]. Steels and its alloys are highly consuming materials by industrialists to produce a variety of structures or products through different manufacturing techniques [2]. Apart from the many advantages of the steels and its alloys, it has some major limitations such as weight, cost, complexities of production, difficulties in creating complex features, lower corrosion resistance in wet environments, etc., which limits its usage for some industrial applications. Because of the mentioned reasons, industrialists/engineers

IntechOpen

looking for another alternative material to replace the steels, found plastics as the best substitute compared to glass and composite materials. Nowadays, plastics/polymers are used to produce a wide variety of industrial products from the very simple ones to the extremely complex featured ones ranging from domestic purpose parts to advanced products such as food storage, medical, optical materials, coatings, electrical devices, electronics, automobiles, space vehicles, etc. [3, 4], due to its good strength to weight ratio, ease of fabrication of complex shapes, low cost, ease of recycling, etc. [5]. Two different types of plastics, such as thermoplastics and thermosets, are extensively used to create industrial products. Thermoplastic materials are softened by applying heat and can be molded into the required shape and size [6, 7]. Important thermoplastics are acrylic, polypropylene, polystyrene, polythene and PVC. Whereas thermosets are formed by heating process and then set its final shape like concrete. The shape of the thermosets cannot modify by reheating them. The materials namely melamine, Bakelite, polyester and epoxy resins are important examples of thermosets. Thermoplastics are the materials which found to be fast growing ones in various industrial purposes due to its excellent properties like strength-to-weight ratio, good high fracture toughness and fatigue resistance, ease of recycling, ease of fabrication of complex parts, low material and processing cost, enhanced thermal insulation and flexibility features [8–11].

The production of complex feature plastics components in any near net manufacturing processes is difficult and not recommended methods regarding to economic aspects. Different types of fabrication methods like plastic adhesive joining, mechanical fastening and welding techniques have been used by various industries [12] to weld two or more plastics parts to create a complete complex plastic product. The welding techniques are more suitable to join plastics compared to other joining techniques, due its advantages like fast and easy processing, tightness of the joint, high strength, etc. [13–15]. There are several welding techniques like tungsten inert gas (TIG) welding, metal inert gas (MIG) and laser welding to join plastics, which are often used by industrialists. Among other methods, laser transmission welding (LTW) is based on lasers and constitutes one of the most innovative fabrication techniques, which are used to join plastic materials into a variety of geometries [16–18] due to process flexibility and better weld quality [19]. Nowadays, the technology of laser transmission welding has been developed to weld a variety of polymers [20] including polycarbonates, acrylic, polyvinylchloride and polyamides using different types of lasers such as CO_2, Nd:YAG, and diode lasers. The usage of LTW in the welding of plastics is increasing due to its advantages and performance capabilities like non-contact, flexible process, easy to control, automate, speed, localized heat, fine spot during welding, etc., compared to other joining technologies.

In LTW, a laser beam is focused on two overlapping thermoplastic materials to be joined. The first part is designed to be transparent to the laser wavelength and second part is absorbent in IR spectrum. The laser beam penetrates the top transparent part and then heat is absorbed and transformed into the underlying absorbing part. Because of the relative motion between the laser beam and the parts to be joined, the laser beam is continuously operating to irradiate the line of specific width of welding zone. This heat is transferred to the transparent part by thermal conduction between the joining parts. Subsequently, surface layers of both materials are melted and welded due to the generated heat. The principle of the laser welding is shown in **Figure 1**. The quality levels of weldment in LTW is depending on the welding material behavior during laser irradiation, controlling of optical properties [21] and correct choice of laser process parameters [22]. To understand the optical phenomena of welding plastic materials various experimental approaches are used to determine the effects of fillers, pigments, reinforcement, combined effect of thickness and color, etc. One of the limitations in laser welding is the energy lost

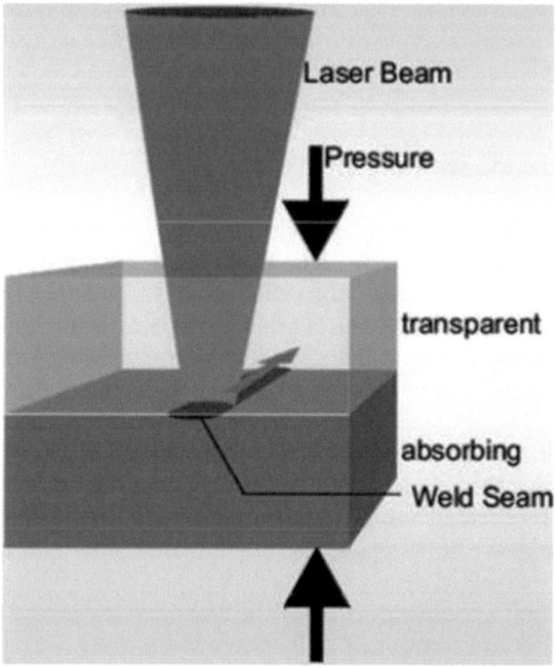

Figure 1.
Schematic diagram showing basic principle of laser transmission welding process, in which laser beam transferred through transparent part to absorbing part and generation of weld seam [34].

due to reflection and absorption due to scattering from the laser transparent part. Polymer degradation and surface burning which affect the strength and the visual appeal of the weldment. In order to achieve a successful weld through laser transmission welding, a good understanding of the issues associated with the welding process are necessary. Due to the complexity of modeling the heating phase of laser transmission welding, experimental methods has been utilized to study welding parameters. In the present chapter, focus has been given to investigate the significances of laser process parameters on weld quality characteristics in laser transmission welding of acrylic materials.

Acrylics are important materials of thermoplastics and its demand is increasing for many industrial applications. Acrylic materials are available in a wide range of thicknesses and colors. They can be opaque, translucent or transparent. They are available in sheet, rod, and tube for use in injection molding, extrusion and vacuum forming. Acrylics withstand all of the weather conditions and it can also be stable in sunlight as well. Transparent acrylics can be used in optical equipment such as cameras due to its finest optical glass property. Acrylics are used widely in typical applications due to their light weight, more impact-resistant than glass, rigidity, good weather resistance, etc. [23]. Various types of acrylics, which are widely used for industrial purposes, include: poly(methyl methacrylate) [24], poly(2-hydroxyethyl acrylate) [25, 26], poly (2-hydroxyethyl methacrylate), etc. [27]. The application of acrylics includes aircraft windows, aerospace [28], electrophoretic applications in the biotechnology industry for separation purposes, identification and preparation of pure samples of nucleic acids, proteins, carbohydrates [29], thermoforming applications such as shower or tub enclosures, outdoor signs [30], medical applications [31, 32], automobiles, house hold applications, etc. For many of the mentioned applications, two or more similar or dissimilar acrylics can be joined to create the complex shape products. Laser transmission welding (LTW) is very often used to join these acrylic materials [23, 33]. LTW of acrylics is an important area of research because of the complexity of the welding process and its associated problems during joining of acrylics [22, 32].

2. Laser transmission welding (LTW)

By means of LTW, welding of plastics is performed by transmitting a laser beam through a laser-transparent material, named as direct laser welding, and weld at interface with the laser-absorbing part, called laser transmission welding [35, 36]. By both welding methods, laser beam is utilized to weld plastic materials. **Figure 2** represents the working principle of laser transmission welding process.

The selection of the laser beam type is depending on the thermo-optical properties of the beam and workpiece. Generally, plastics are transparent to laser radiation at its nature state. This property helps to weld the plastics very easily. During welding, when laser beam focused on transparent plastic sheet, some fractions of the incident light is reflected and the remaining light is transmitted through the job. The absorbing capability of the plastic part by radiation energy can be calculated by the Beer–Lambert law. Molecular diffusion can occur and with the application of clamping pressure, a contact portion of the two parts are joined and thus, a solid joint is formed. Different steps of LTW is shown in **Figure 3**.

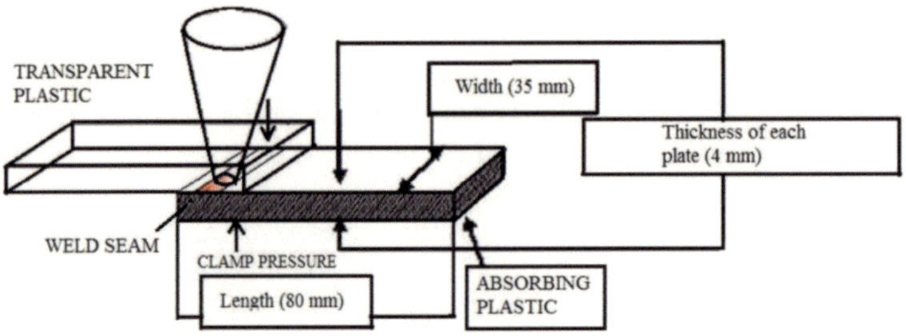

Figure 2.
Schematic diagram showing laser transmission welding setup in which one can find both transparent plate and absorbing plate and focused laser beam and weld seam generation [37].

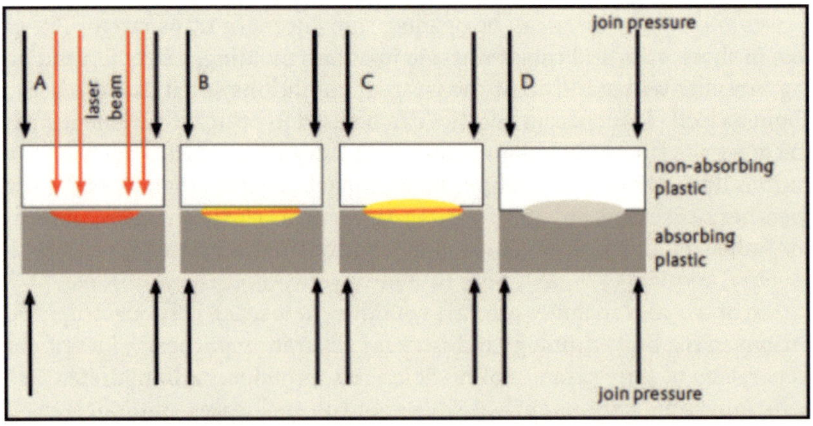

Figure 3.
Schematic diagram showing three stages of laser welding process to join plastic materials with the application of laser beam, melting and welding of plastic plates [38].

3. Literature survey

Very few research studies on LTW of plastics have been done. Thus, several aspects need to be extensively explored to enhance the welding performance. In the present study, transmission laser welding of thermoplastics, especially acrylics, has been analyzed with respect to varying input parameters. Some of the reported articles from the literature related to welding of acrylics using LTW are discussed in the following paragraphs.

Mandal et al. [23] maximized the weld strength of laser beam welding of transparent and opaque acrylic specimens. Researchers considered clamping pressure, scanning speed and current as process parameters. Thus, they found from their investigation that the strength of the welded joint was highly influenced by the variation of scanning speed, current flow and clamping pressure. On the other hand, Acherjee et al. [33] studied and analyzed the laser transmission welding of acrylics to determine the optimal welding parameters to attain a desired weld seam with maximum joint strength. Direct and interaction effects of process welding parameters on output responses were illustrated through overlay contour plots. The optimal parametric setting was obtained by overlay contour plots to enhance the welding performance. Acherjee et al. [34] also carried out research work on laser transmission welding process of thermoplastics (acrylics) to predict the multiple quality characteristics. They applied combined applications of gray-Taguchi methodologies to plan the experiments and optimize the multi-responses, weld strength and weld width by controlling welding parameters, like laser power, welding speed and defocal position. Analysis of variance was applied on experimental data to identify the most significant process variable for the overall output feature of the LTW operation. They stated from ANOVA results that laser power had the most dominant effect on both responses: defocal position and welding speed.

Kumar et al. [37] studied and analyzed the influences of laser process parameters to optimize weldment quality performances by through transmission laser welding (TTLW) of acrylic materials. They conducted the experiments based on orthogonal array of Taguchi method to determine the factor effects on output responses of acrylic joints by TTLW. The statistical tools like signal-to-noise ratio and analysis variance were utilized to depict the significances of laser process variables on joint strength and weld width of laser welded joint. From the analysis, researchers stated that quality characteristics of laser-welded acrylics were significantly influenced by process parameters and the Taguchi method was robust tool for analyzing and optimizing the TTLW of acrylics. Acherjee et al. [39] studied and optimized the weld quality characteristics by performing experiments on joining acrylics (poly(methyl methacrylate)) specimens by laser transmission welding by varying process parameters namely laser power, welding speed, size of the laser beam and clamp pressure. They developed the mathematical models between the input parameters and out responses to generate mathematical relationships between them. Analysis of variance (ANOVA) had been used to determine the effects of laser variables on weldment responses: lap shear strength and weld seam width. Investigators stated from the performed research that the control and the optimization of the process parameters were very important in laser transmission welding to enhance its efficiency. Acherjee et al. [40] utilized the integrated Taguchi method and desirability function approach to analyze and optimize the welded joint responses of acrylic plastic material in LTW. The authors conducted experiments as per orthogonal array of Taguchi method. ANOVA was used to find the influential process variables on weld responses. From this study, the researchers mentioned that Taguchi method and ANOVA were useful tools to study and optimize the process conditions to improve the responses of acrylic welded joints in LTW.

Acherjee et al. [41] carried out an experimental investigation to analyze the significances of process variables and control them to enhance the process efficiency of LTW of acrylic materials. Experimental runs were performed by orthogonal array of Taguchi method. The dependency of weld strength on correct selection welding parametric conditions were studied by using the analysis variance technique. The authors of this work stated from the study that laser power showed the highest influence on weld strength, followed by focal distance and welding speed. The optimal parametric setting was obtained by Taguchi method, which was conformed through the confirmatory test. Barma et al. [42] conducted experimental analysis to investigate the weld joint characteristics of acrylic components under varying input process conditions namely scanning speed, current and clamping pressure. These investigators studied the variations of weld quality responses with varying process parameters. They evaluated some parametric combinations to obtain desired weld quality. Nitesh et al. [43] carried out a research to analyze, model and optimize the laser process variables namely laser power, scanning speed and frequency to improve the quality characteristics of acrylic and polycarbonate welded joint in through transmission laser welding. Analysis of variance was utilized to determine the influential process parameters, which had detrimental effects on process performances. They stated from this study that selection of optimal parametric setting plays a critical role to obtain the desired weld joint qualities.

Kumar et al. [44] studied the effects of process parameters on the circular contour laser transmission welding of acrylic plastics. Experiments were conducted based on orthogonal array of Taguchi method by considering laser current, stand-off distance and rotational speed as input parameters. The significances of process parameters on output responses, weld width and breaking load were studied by ANOVA. They mentioned in their research that welding responses were significantly influenced by process parametric conditions. Acherjee et al. [45] made an experimental investigation on diode laser transmission welding of dissimilar thermoplastics between poly(methyl methacrylate) and acrylonitrile butadiene styrene to study the effects of the laser welding parameters like laser power, welding speed, stand-off distance and clamp pressure on responses: weld strength and weld width. The ANOVA technique was applied on experimental data to identify the significant process variables, which were expected to influence the welding responses. Graphical contour plots were also drawn to study the direct and interaction effects of input variables on welding performances. Nakhaei et al. [46] utilized the applications of artificial neural network (ANN) to study and model the effects of laser power, welding speed, clamp pressure and stand-off distance on output response, weld lap-shear strength in laser transmission welding (LTW) of acrylics. They developed ANN based model to analyze and predict the complex relationships between laser parameters and weld quality response. They observed from their research analysis that weld lap-shear strength was increasing with increasing laser power and clamp pressure and decreasing with increasing welding speed and stand–off distance.

Kumar and Bandyopadhyay [47] analyzed the contour laser transmission welding (LTW) of thermoplastics by process modeling, finite element method (FEM) and response surface methodology (RSM) integrated approach. They studied the effect of process parameters on temperature field and weld bead dimension. Interaction effects of process parameters on the responses were studied using the graphical response surface plots. They observed from their investigation that responses such as weld width, depth of penetration in absorbent part and maximum temperature at weld interface decreased with increasing scanning speed, laser power and spot diameter, which were significant factors on the maximum temperature at weld interface. Seo et al. [48] used LTW to join acrylics and other thermoplastic materials. These experiments were conducted with varying specific energy and weld time. They stated that LTW was not only a useful method for welding acrylics but also cost effective in comparison with the other plastics joining methodologies. Shashi et al. [49] made a study on pulsed Nd:YAG laser welding of poly(methyl methacrylate) to determine the influences of process parameters namely lamp current, pulse frequency, pulse width and cutting speed, on micro-channel characteristics, microchannel width, microchannel depth, burr width and burr height. They conducted experiments based on response surface methodology (RSM). Mathematical relationships between the input parameters and output responses were made by RSM. They explored the variations of output responses with varying input parameters using ANOVA and graphical plots and stated that process parameters were most significant for enhancing the welding performance.

From the extensive literature survey made in the present chapter, it can be stated that acrylics are the one of the most important industrial material family of polymers. The usage of these materials is growing for various industrial needs as mentioned earlier [50–53]. Laser transmission welding (LTW) is effective and efficient for welding of acrylics or other polymer materials [54–59]. Parametric studies related to various aspects of joining of acrylics by LTW can make sound knowledge base, from which industrial persons may select optimal parametric settings to produce economically and predictably desired weld joint quality performances. The present chapter is one-step forward for it. Many more related studies will make the LTW of acrylics simple and easy.

4. Conclusions

The conclusions drawn from the present study of joining of acrylic materials using the laser transmission welding (LTW) process are presented as follows:

i. Importance and necessity of welding of plastics have been discussed.

ii. Thermoplastic materials are extensively used in advanced industrial purposes such as automotive, aerospace, medical, etc.

iii. The importance of joining of thermoplastics, especially acrylics, have been discussed.

iv. Laser transmission welding is one of the most important plastic joining techniques, among others, which is used to create complex shape jobs.

v. Understanding the relationships between input and output responses plays a critical role in welding of acrylics by LTW.

vi. Correct selection of laser process parameters is crucial to obtained desired weld qualities on acrylic welded joints.

vii. Polymer degradation and surface burning are notable defects during thermoplastic welding.

viii. The effects of laser parameters on performance characteristics of weldments needs to be properly understood to conduct LTW in an optimum manner.

ix. Statistical tools like analysis of variance, signal-to-noise ratio are useful to study and analyze the factor effects on outcome of the LTW.

x. More experimental research works related to thermoplastic welding by laser transmission welding (LTW) needs to be conducted to enhance the performance of welding.

Acknowledgements

The author acknowledges to Dr. Asish Bandyopadhyay, Professor, Mechanical Engineering Department, Jadavpur University, India, for encouraging me to write this chapter.

Conflict of interest

The author declaring no conflict of interest.

Author details

Ramesh Rudrapati
Department of Mechanical Engineering, Hawassa University—Institute of Technology, Hawassa, Ethiopia

*Address all correspondence to: rameshrudrapati@gmail.com

References

[1] Chattopadhyay AB. Machining and Machine Tools. Delhi, India: Wiley; 2011

[2] Rudrapati R. Machining of stainless steels and alloys using non-traditional machining processes. In: Stainless Steels and Alloys. London, UK: Intech Open; 2018. pp. 15-30

[3] Reitz WE, Rebecca MO. How to join plastics. Advanced Materials and Processes. 2000;**158**(3):49-52

[4] Vendan SA, Natesh M, Garg A, Gao L. Polymer welding techniques and its evolution. In: Confluence of Multidisciplinary Sciences for Polymer Joining. 2019. pp. 11-71

[5] Genna S, Tagliaferri F, Papa I, Leone C, Palumbo B. Multi-response optimization of CFRP laser milling process based on response surface methodology. Polymer Engineering and Science. 2017;**57**(6):595-605

[6] Tsao CW, DeVoe DL. Bonding of thermoplastic polymer micro fluidics. Micro Fluid Nano Fluid. 2009;**6**:l-16

[7] Haghshenas M, Khodabakhshi F. Dissimilar friction-stir welding of aluminum and polymer: A review. International Journal of Advanced Manufacturing Technology. 2019. DOI: 10.1007/s00170-019-03880-2

[8] Patrick L, Volker S. Simulation-based investigation on the temperature influence in laser transmission welding of thermoplastics. Welding in the World. 2019;**63**(2):221-228

[9] Becker H, Gartner C. Polymer micro fabrication technologies for micro fluidic systems. Analytical and Bioanalytical Chemistry. 2008;**390**:89-lll

[10] Marczis B, Czigany T. Interrelationships between welding parameters of hot-gas welded polypropylene. Polymer Engineering & Science. 2006;**46**(9):1173-1181

[11] Balkan O, Halil D, Ayhan E, Hüseyin Y. Effects of welding procedures on mechanical and morphological properties of hot gas butt welded PE, PP, and PVC sheets. Polymer Engineering & Science. 2008;**48**(4):732-746

[12] Dodin MG. Welding mechanisms of plastics: A review. The Journal of Adhesion. 1981;**12**(2):99-111

[13] Herfurth H, Ehlers B, Heineman S, Haensch D. New approaches in plastic welding with diode lasers. In: Proceedings of the 18th International Congress on Applications of Lasers and Electro-Optics, LMP Section B. San Diego, CA, USA; 1999. pp. 48-56

[14] Zak G, Mayboudi L, Chen M, Bates PJ, Birk M. Weld line transverse energy density distribution measurement in laser transmission welding of thermoplastics. Journal of Material Processing Technology. 2010;**210**:24-31

[15] Amanat N, James NL, McKenzie DR. Welding methods for joining thermoplastic polymers for the hermetic enclosure of medical devices. Medical Engineering & Physics. 2010;**32**:690-699

[16] Pardeep P, Avinish T, Pankaj B. Transient thermal analysis of CO2 laser welding of AISI 304 stainless steel thin plates. In: Manufacturing Engineering. 2019. pp. 49-66

[17] Mishra D, Sahu SK, Mahto RP, Pal SK, Pal K. Friction stir welding for joining of polymers. In: Strengthening and Joining by Plastic Deformation. 2019. pp. 123-162

[18] Kagan V, Pinho G. Laser transmission welding of semicrystalline thermoplastics-part II: Analysis of mechanical performance of welded nylon. Journal of Reinforced Plastics and Composites. 2004;**23**(1):95-107

[19] Gupta SK, Pal PK. Analysis of through transmission laser welding of nylon6 by finite element simulation. Management and Production Engineering Review. 2018;**9**:56-69

[20] Navas-Martos FJ, Yebra-Rodríguez A, Dolores LRM. Laser transmission welding of poly (lactic acid) and polyamide66/sepiolite nanocomposites. Journal of Applied Polymer Science. 2018;**135**:46638

[21] Geiger M, Frick T, Schmidt M. Optical properties of plastics and their role for the modeling of the laser transmission welding process. Production Engineering. 2009;**3**:49-55

[22] Rudrapati R, Kumar N, Pal PK. Applications of Taguchi method for parametric optimization of through transmission laser welding of plastics. AIP Conference Proceedings. 2019;**2057**:020013-1-020013-7

[23] Mandal RK, Bandyopadhyay A, Santanu D. An experimental investigation on laser beam welding of acrylics. Indian Welding Journal. 2014;**47**:75-80

[24] Serrano-ArocaÁ, Pradas-MonleónM, Ribelles-Gómez JL. Macroporous poly(methyl methacrylate) produced by phase separation during polymerisation in solution. Colloid and Polymer Science. 2007;**285**(7):753-760

[25] Monleón-PradasM, Gómez-RibellesJL, Serrano-Aroca A, Ferrer-Gallego G, Anton JS, Pissis P. Interaction between water and ploymer chains in poly (hydroxyethyl acrylate) hydrogelas. Colloid and Polymer Science. 2001;**279**:323-330

[26] Serrano-Aroca A, Pradas MM, Ribelles JLG, Rault J. Thermal analysis of water in reinforced plasma-polymerised poly (2-hydroxyethyl acrylate) hydrogels. European Polymer Journal. 2015;**72**:523-534

[27] Ramaraj B, Ganga R. Modification of the dynamic swelling behaviour of poly (2-hydroxyethyl methacrylate) hydrogels in water through interpenetrating polymer networks (IPNs). Polymer. 1994;**35**(10):2167-2173

[28] Kanwal JS, Jatinder K. Chemical assisted USM of acrylic heat resistant glass. In: Manufacturing Engineering. 2019. pp. 167-184

[29] Soane DS, Soane ZM. Acrylic microchannels and their use in electrophoretic applications. US Patent 6, 054, 034

[30] Clark TH. ABS/Acrylic lamination process. US Patent 5, 069, 851

[31] Bożena T, Anna D, Sonia KK. Acrylates in dental applications. In: Acrylic Polymers in Healthcare. 2017. pp. 25-42

[32] Brígido-Diego R, Pérez-Olmedilla M, Serrano-Aroca A, Gómez-Ribelles JL, Pradas-Monleón M, Gallego-Ferrer G, et al. Acrylic scaffolds with interconnected spherical pores and controlled hydrophilicity for tissue engineering. Journal of Materials Science: Materials in Medicine. 2005;**16**:693-698

[33] Acherjee B, Misra D, Bose D, Acharyya S. Optimal process design for laser transmission welding of acrylics using desirability function analysis and overlay contour plots. International Journal of Manufacturing Research. 2011;**6**:49-61

[34] Acherjee B, Arunanshu SK, Souren M, Dipten M. Application of grey-based Taguchi method

for simultaneous optimization of multiple quality characteristics in laser transmission welding process of thermoplastics. International Journal of Advanced Manufacturing Technology. 2011;**56**:995-1006

[35] Mahmood T, Mian A, Amin MR, Auner G, Witte R, Herfurth H, et al. Finite element modeling of transmission laser microjoining process. Journal of Material Processing Technology. 2007;**186**(1-3):37-44

[36] Kumar N, Sherlock R, Tormey D. Prediction of weld interface depth and width at optimum laser welding temperature for polypropylene. Procedia CIRP. 2019;**81**:1272-1277

[37] Kumar N, Rudrapati R, Pal PK. Multi-response optimization in through transmission laser welding of acrylic plastics using grey based Taguchi method. Procedia Materials Science. 2014;**5**:2178-2187

[38] Van de Ven JD, Erdman AG. Laser transmission welding of thermoplastic - part I: Temperature and pressure modeling. Journal of Manufacturing Science and Engineering. 2007;**129**:849-858

[39] Acherjee B, Misra D, Bose D, Venkadeshwaran K. Prediction of weld strength and seam width for laser transmission welding of thermoplastic using response surface methodology. Optics & Laser Technology. 2009;**41**(8):956-967

[40] Acherjee B, Kuar AS, Mitra S, Misra D. Integrating Taguchi method with desirability function analysis to optimize the laser transmission welding process. American Institute of Physics. 2010;**1298**:480-485

[41] Acherjee B, Kuar AS, Mitra S, Misra D. Selection of process parameters for optimizing the weld strength in laser transmission welding of acrylics. Proceedings of the Institution of Mechanical Engineers, Part B. 2010;**224**:1529-1536

[42] Barma JD, Pal PK, Bandyopadhyay A. Study on the effects of different process parameters in through transmission laser welding. In: Proceedings of the Seminar on Application of Laser in Materials Processing. Jadavpur, Kolkata; 2010. pp. 193-195

[43] Nitesh K, Nikhil K, Bandyopadhyay A. Optimization of pulsed Nd: YVO4 through transmission laser welding of transparent acrylic and polycarbonate. Materials Today: Proceedings. 2018;**5**: 5235-5243

[44] KumarN,MondalS,BandyopadhyayA, Bhattacharyya BK. Effects of process parameters on the circular contour laser transmission welding process. Lasers in Engineering. 2017;**36**:299-312

[45] Acherjee B, Arunanshu SK, Souren M, Dipten M, Sanjib A. Experimental investigation on laser transmission welding of PMMA to ABS via response surface modeling. Optics & Laser Technology. 2012;**44**:1372-1383

[46] Nakhaei MR, Mostafa ANB, Kordestani F. Modeling of weld lap-shear strength for laser transmission welding of thermoplastic using artificial neural network. Advanced Materials Research. 2012;**445**:454-459

[47] Kumar N, Bandyopadhyay A. Simulation of the effects of input parameters on weld quality in laser transmission welding (LTW) using a combined response surface methodology (RSM)-finite element method (FEM) approach. Lasers in Engineering. 2017;**36**:225-243

[48] Seo M, Ryu K, Nam G. Study on the laser transmission welding of thermoplastics. Journal of Korean Society of Precision Engineering. 2005;**22**(9):34-40

[49] Shashi P, Acherjee B, Arunanshu SK, Souren M. An experimental investigation on Nd:YAG laser microchanneling on polymethyl methacrylate submerged in water. Proceedings of the Institution of Mechanical Engineers, Part B: Journal of Engineering Manufacture. 2013;**227**:508-519

[50] Rahmah M, Noraspalelawati R, Abang AE, Sahbudin S. Characterisation and process optimisation of photosensitive acrylates for photonics applications. Science and Technology of Advanced Materials. 2005;**6**:375-382

[51] Linval RDP, Robert RM, Carrol SW, Flo W. Dermal oncogenicity bioassays of monofunctional and multifunctional acrylates and acrylate-based oligomers. Journal of Toxicology and Environmental Health. 1985;**16**:55-60

[52] Jouni S, Tomas L, Gunnar R, Torkel F. Surface contamination to UV-curable acrylates in the furniture and parquet industry. Applied Occupational and Environmental Hygiene. 2001;**16**:360-368

[53] John DH, Kevin HR, James EM. Aquatic risk assessment of acrylates and methacrylates in household consumer products reaching municipal wastewater treatment plants. Environmental Technology. 1995;**16**(8):715-727

[54] Liu HX, Yan Z, Li P, Wang X. Prediction of molten area in laser transmission welding of thermoplastic polymers. Science and Technology of Welding and Joining. 2014;**19**:487-492

[55] Mayboudi LS, Birk AM, Zak G, Bates PJ. A three-dimensional thermal finite element model of laser transmission welding for lap-joint. International Journal of Modelling and Simulation. 2009;**29**:149-155

[56] Moskvitin GV, Polyakov AN, Birger EM. Laser welding of plastics (review). Welding International. 2013;**27**(9):725-734

[57] Devrient M, Frick T, Schmidt M. Laser transmission welding of optical transparent thermoplastics. Physics Procedia. 2011;**12**:157-165

[58] Val AK. Innovations in laser welding of thermoplastics: This advanced technology is ready to be commercialized. Journal of Materials and Manufacturing. 2002;**111**:845-864

[59] Grewell D, Rooney P, Kagan VA. Relationship between optical properties and optimized processing parameters for through-transmission laser welding of thermoplastics. Journal of Reinforced Plastics and Composites. 2004;**23**:239-247

Chapter 3

Properties and Applications of Acrylates

Kingsley Kema Ajekwene

Abstract

Acrylates are the esters, salts and conjugate bases of acrylic acid with its derivatives. They are made from acrylate monomer, which usually comprises of esters which contains vinyl groups, that is two carbon atoms that are double-bonded to each other, and directly attached to the carbonyl carbon of the ester group. Acrylates possess very diverse characteristic properties ranging from super-absorbency, transparency, flexibility, toughness and hardness, among others. These kinds of materials are used in sundry applications such as diapers, cosmetics, orthopedics, paints and coatings, adhesives, textiles, and many biomedical applications such as contact lenses and bone cements. This book chapter highlights the characteristic properties and applications of acrylates, its derivatives and copolymers.

Keywords: acrylates, methacrylates, super-absorbency, transparency, coatings, properties, applications

1. Introduction

Acrylates (or polyacrylates), belong to the family of vinyl polymers made from acrylate monomers. They are the esters, salts and conjugate bases of acrylic acid and its derivatives, of which methacrylates are the most common members. Its monomers are esters containing vinyl groups, which comprises two carbon atoms that are double-bonded to each other and directly attached to the carbonyl carbon of the ester group. This makes acrylates to be bifunctional in nature: the vinyl group is susceptible to polymerization, hence; carbonyl group carries myriad functionality using alcohols or amines [1–9]. The resulting polymers are characterized with a high molecular weight with physical and chemical properties that depend on the lateral substituents of the polymeric chains [10].

Acrylates are important class of polymers that are rubbery, soft and tough. They are well-known for their good impact toughness and resistance to breakage, transparency, elasticity, fairly good heat and oil resistance. They have also the capacity to endure a temperature of 170–180°C when they are in oil, exposed to dry heat or non-surfactant suspending agents that have anti-static binding abilities and film-forming. They also have good weatherability and ozone resistance due to the absence of double bonds in their backbone. Their glass transition temperature (T_g) is usually well below the room temperature [1–4].

Acrylate polymers are commonly used in dentistry and many other biomedical applications [11], cosmetics and artificial nail products such as eyelashes, nail enhancing polishes, nail builders, artificial nails and to help artificial nails mold to the natural nail plate as an adhesive, hair fixatives, in marine phytoplankton as a poisonous defense against predators such as protozoa, and oil seals and packaging that are related to automobiles [1–4, 8].

Acrylate monomers include ethyl acrylate, ethylene-methyl acrylate, methyl methacrylate, 2-chloroethyl vinyl ether, 2-hydroxyethyl acrylate, hydroxyethyl methacrylate, butyl acrylate, trimethylolpropane triacrylate (TMPTA).

The preparation of acrylate polymers usually involves the treatment of acrylic acids with the corresponding alcohol in the presence of a catalyst [1–4].

2. Properties of different types of acrylates

2.1 Polyacrylic acid

Polyacrylic acid (or 2-propenoic acid) is the simplest acrylate polymer $(C_3H_4O_2)_n$.

Polyacrylic acid polymers are usually produced from acrylic acid that is obtained from propene (a byproduct of ethylene and gasoline production). Acrylic acid polymer is used as thickening agent in acidic conditions because it is highly water dispersible and its dispersion in water thickens with the increase of pH [5].

Polyacrylic acid and related acrylates are regarded as unusual polymers because they absorb water very freely. They soak up water in multiples of their own weight with no difficulty. This means that even as a little grams of the polymer can absorb a cup of liquid, lock it in and hold it up tightly. Such polymers with this characteristics are termed as "super-absorbers". Therefore, it has found usefulness in diapers, especially baby diapers. In diapers, polyacrylic acid absorbs all the liquids that little babies excrete and deposit in it. The super-absorbency of polyacrylics is advantageous in that way that they are less messy and more hygienic because the mess is locked up in the polyacrylic acid polymer and babies feel comfortable while their parents take their time to change the diapers [1–4]. That is why babies that use diapers of this sort do not get unpleasant skin rashes compared to those using napkins.

Acrylic acids form corresponding esters when reacted in the presence of an alcohol. Acrylic acid and its associated acrylates play important roles in the manufacturing of plastics, diapers, adhesives, floor polishes, nail varnishes/polish, paints, coatings and adhesives of various types [12].

With its film-forming characteristics, it serves as a film-forming agent that is used to obtain continuous films on the nail, skin or hair. These characteristics enable them to be used in various applications including eyeliner, liquid makeups, mascaras, nail polish, sunscreens, lipsticks and skin care products (creams, lotions, etc.).

They also find major applications in coatings and paints such as solvent-born coatings, emulsion paints, interior and exterior water-based paints, and printing inks for applications that need quick drying rates such as auto based lacquers and industrial coatings. They are also used in pressure sensitive adhesive formulations—from low adhesion that are barely tacky to very high tack that can bond permanently to substrates, textiles, automotive products, leather finishing, tape adhesives, high-temperature-resistant and oil-resistant elastomers. They can also be used as plasticizer to improve the plasticity of brittle and rigid plastics that are largely used in the automotive industry for dampers, hoses, gaskets and seals that usually function under long-term exposure to hydrocarbon oils and elevated temperatures [1–4].

2.2 Methacrylate

Methacrylate is often used as a generic term for methacrylate and acrylate. Methacrylate refers to derivatives of methacrylic acid or methyl ester of acrylic acid.

These derivatives include the acid form, $CH_2C(CH_3)CO_2H$ or ($C_4H_6O_2$), the salts such as $CH_2C(CH_3)CO-_2$ Na+ and the esters (e.g. $CH_2C(CH_3)CO_2CH_3$ or methyl methacrylate) and their corresponding polymers.

Methacrylates can form polymers easily because their double bonds are usually very reactive. The polymers exhibit a characteristic water solubility which is pH-dependent. This behavior makes them useful as coating agents in drug delivery systems. It is mostly used as a vehicle for obtaining amorphous systems and solid dispersions, such as nanoparticles, microparticles and microspheres to increase the solubility of low solubility drugs [1–4, 12, 13].

Poly(methacrylate) hydrogels have found useful applications in medical devices, especially for ocular applications (e.g. intraocular lenses and contact lenses), and as cell delivery and drug delivery systems. They have found potential applications as intelligent hydrogels for biomimetic sensors, controlled drug release, analyte-sensitive materials, intelligent polymeric membranes and micro-fluidic devices [1].

2.3 Poly(methyl methacrylate)

Poly(methyl acrylate) (PMA) is produced with the methyl acrylate (MA) monomer, also called as methyl prop-2-enoate, methyl propenoate or methoxy-carbonylethylene. However, poly(methyl methacrylate) (PMMA) is synthesized by polymerization of methyl methacrylate (MMA), which is an acrylate monomer in which one carboxylic acid hydrogen and vinyl hydrogen are replaced by a methyl group.

Methylacrylate is produced by the reaction of esterification with methanol under acid catalysis (acidic ion exchangers, p-toluene sulfonic acid, sulfuric acid) or by pyrolysis of methyl lactate in the presence of ethenone (ketene), the debromination of methyl 2,3-dibromopropanoate with zinc or dehydration of methyl lactate over zeolites as well as vapor-phase oxidation of nickel tetracarbonyl-catalyzed hydro-carboxylation of acetylene and propene or 2-propenal with oxygen in the presence of methanol with carbon monoxide which also yields methyl acrylate. Reacting methyl formats with acetylene in the presence of a transition metal catalyst can also yield methyl acrylate [1–4, 9, 13].

PMMA is a linear chain polymer that is characterized by its hydrophobicity, relatively superior light admittance (transparency), amorphous disposition at a molecular level, and glassy feature at room temperature that makes it a strong, hard and clear plastic under the "Plexiglas" trade name [10]. However, it can also become a white rubber at room temperature. The hardness or softness is determined by the numbers of methyl groups present which determines chain mobility and hence the softness and flexibility (i.e., the extent of the free movement of the polymer chains around each other). The softer the polymer, the better the mobility and pliability. Therefore, a polymer will be soft or hard depending on its easy or difficult movement.

MMA is used as the monomer resin in dental materials, in windscreen repair kits and as bone cement for fixing prosthetic devices in orthopedic surgery where it provides a buffering effect, absorbing shock from forces acting upon the bone because of its optimal ability to efficiently distribute stresses and interface strain energy [10]. In its pre-processed form, it is usually a colorless liquid characterized by an acrid odor. It is mainly produced to make acrylate fiber, which is used in weaving synthetic carpets. It also serves as a reagent in the synthesis of various pharmaceutical intermediates. PMMA is used in the preparation of 2-dimethylaminoethyl acrylate by trans-esterification with dimethyl amino-ethanol. It is used as a co-monomer with a variety of vinyl and acrylic monomers in the polymerization of various polymers. When using MA as co-monomer, the resulting acrylates are more brittle and harder than those with homologous acrylates. It is also a good dienophile.

PMMA is a non-degradable polyacrylate, which is the most commonly utilized non-metallic implant material in orthopedic surgery and it is also used as an essential ingredient in making dentures. PMMA can be applied as a bone fixing material in the implanting of orthopedic prosthetic materials for shoulders, knees and hips repairs. PMMA-based bone cements can be added to ceramics or bioactive glasses to impart mechanical properties and modify the curing kinetics and as well as loading some dosages of antibiotics in the PMMA cement to decrease the risk of related prosthesis infection. Due to its outstanding bio- and hemo-compatibility and ease of modification, PMMA is usually used in several medical devices, including dialyzers and blood pumps. The optical properties make it a good material for hard contact lenses and implantable ocular lenses. PMMA also offers coloring and physical properties that are beneficial for denture fabrication [14].

PMMA can be introduced into cosmetic products as beads (porous and non-porous beads) to enhance its performance [13, 15]. The non-porous beads of PMMA impart very high resistance to organic solvents, which allows its usage in nail varnish (mattifying and hardening effects). It has a supple structure and low oil absorption making it a good binder for pressed powders. In addition, they are absolutely transparent on the skin top giving rise to an excellent volume in mascaras, velvet touch effect and ball-bearing effect. Due to their non-porous nature, they do not absorb moisture and sebum neither do they dry the skin. Thus, it makes them ideal candidates for normal, dry and sensitive skin. They are versatile compounds, which can provide soft feel on the skin, optical effects and unique textures to achieve innovative formulations for skin care and make-up [15].

The porous beads of PMMA are cross-linked and have very high specific area that facilitates the enhancement of absorption of sebum and sweat, occasioning a matte and dry influence on the skin (preventing the sticky skin feel after the application) and optically suppressing makeup smudging [15].

2.4 Polyacrylamide

Polyacrylamide (PAM) readily dissolves in water. Therefore, this polymer is used industrially in many applications where it serves such purposes. In addition, the cross-linked form of polyacrylamides can also absorb water to form gels without dissolving. The absorbed water in the formed gels of cross-linked polymer renders them soft and thus suitable for use in biomedical products such as soft contact lenses [1–4].

Another application of PAM gels include soil conditioner to stabilize soil aggregates as well as flocculate suspended particles and furrow irrigation where it reduces erosion and runoff while improving soil and water quality and water-use efficiency. They are also used to reduce surface sealing and crusting as well as erosion in certain agriculture and sprinkler irrigation systems in a cost effective way. PMA is also used to stabilize steep slopes in construction, highway cuts, and other disturbed soils in geo-construction [16].

2.5 Polyacrylonitrile

Polyacrylonitrile (PAN) is a derivative of acrylic acid where the carboxylic acid group is replaced by the related nitrile group and presents the molecular formula $(C_3H_3N)_n$.

PAN is a semi-crystalline polymer and though it is a thermoplastic, it does not melt under normal conditions. Its degradation occurs before it melts and its melting is usually observed at above 300°C. When this polymer is heated up to 200°C, it can be transform into a rigid structure with a release of energy, an occurrence referred to as cyclization. Above the cyclization temperature, this polymer undergoes oxidation, a scenario that confers non-flammability on it. If PAN is heated to above 1000°C under inert atmosphere, it gives a product with a percentage of carbon greater than 90%. This property makes it usable in carbon fiber production.

Some major characteristic properties that render PAN special compared to other polymers include the fact that it is the most resistant polymer to degradation by ultraviolet rays and sunlight and it is highly chemically unreactive and resistant to most organic acids and solvents. In addition, this polymer can be attacked only by concentrated solutions of alkalis and highly polar liquids. PAN in the form of acrylic fiber is resistant to breakages, is soft and comfortable, possesses high thermal insulation properties, and it does not melt maintaining its morphological structure when it is heated. This last property of PAN is utilized for the production of insulation fibers, flame retardant fibers, carbon fiber and blankets for filtration of hot gases [17].

Due to its special properties such as stability to UV degradation, low density, high strength and modulus of elasticity, thermal stability, non-fusibility and chemical resistance, PAN is an important polymer for a wide range of applications. Thus, it is used in textile applications due to its wool-like characteristics, in the making of carbon fiber, filtration membranes, fibers for cement reinforcement, sails for yachts, awning fabrics in outdoor applications, specialist fibers for acoustic and thermal insulation, anti-flame fibers and in the production of felts for hot air filtration systems [17, 18].

Copolymers containing polyacrylonitrile, which are flame-retardant, are used as fibers to make knitted clothing, like socks and sweaters, as well as outdoor products like tents. Poly(acrylonitrile-*co*- butadiene-*co*--styrene) (ABS) and Poly(styrene-*co*-acrylonitrile) (SAN) are used as plastics. ABS is a lightweight and very strong polymer. It is so light and strong enough to be used in the production of automobile body parts. Thus, this material render automobiles' lighter using less fuel and thus polluting less [1–4].

The strength of ABS emanates from the nitrile groups that are present in its acrylonitrile units. The nitrile groups are very polar; hence, they can be attracted to one another. This occurrence permits opposite charges on the nitrile groups to stabilize one another. This strong attraction holds ABS chains very tightly, thus, it makes the material very strong. Furthermore, the rubbery polybutadiene present in it makes ABS a tough polymer [1–4, 19].

2.6 Ethyl acrylate

Ethyl acrylate (EA), also named as ethyl propenoate, ethyl prop-2-enoate, acrylic acid ethyl ester or ethyl propenoate, presents the molecular formula $(C_5H_8O_2)_n$.

Ethyl acrylate is the ethyl ester of acrylic acid. It is a colorless liquid with a distinctive unpleasantly pungent odor and it is mainly used in the production of paints and textiles, especially non-woven fibers. EA can also be used as a reagent during the synthesis of various pharmaceutical intermediates.

Ethyl acrylate is prepared by acid-catalyzed esterification of acrylic acid that is usually produced by oxidation of propylene. It can also be prepared from ethanol, carbon monoxide and acetylene by a Reppe reaction.

EA is used in the manufacture of polymers including rubber, plastics, denture material and resins. It is used as a reactant for alkyl acrylates (acrylic esters) by trans-esterification with higher alcohols through basic or acidic catalysis.

Other acrylate monomers: 2-ethyl hexyl acrylate (obtained from 2-ethylhexanol) is used in pressure-sensitive adhesives; cyclohexyl acrylate (obtained from cyclohexanol) is used in automotive clear lacquers; 2-hydroxyethyl acrylate (obtained from ethylene glycol), which is crosslinkable with di-isocyanates to form gels, is used with long-chain acrylates as co-monomer in comb polymers for the reduction of the solidification point of paraffin oils; 2-dimethylaminoethyl acrylate (obtained from dimethylaminoethanol) is used in the preparation of flocculants for sewage clarification and production of paper [8, 13].

Ethyl acrylate is a reactive monomer. Therefore, it is useful in the enhancement of the cleaning effect of liquid detergents, used as gastric juices tablet covers, in paper, paints and adhesives, textile and leather auxiliaries together with cosmetic and several other pharmaceutical products. EA is also used as a flavoring agent.

2.7 2-Ethyl hexyl acrylate

2-Ethyl hexyl acrylate, also named as 2-ethyl hexyl prop-2-enoate or abbreviated as 2-EHA with the molecular formula $(C_{11}H_{20}O_2)_n$, is a chemical compound that belongs to the acrylates' family. It is a colorless liquid, which has a sweet odor and can be prepared by the esterification of racemic 2-ethyl hexanol and acrylic acid in the presence of the hydroquinone polymerization inhibitor with a strong acid such as methane-sulfonic acid by reactive distillation using the toluene azeotroping agent.

2-Ethyl hexyl acrylate easily and readily polymerizes. The polymerization of this monomer can be initiated by heat, peroxides, light or even by contaminants. In the presence of strong oxidants, 2-EHA reacts violently with the tendency to form explosive mixtures when exposed to air at temperatures above 82°C (180°F). The chemical, physical and toxicological properties, however, can be greatly modified by additives or stabilizers. 2-Ethylhexyl acrylate is one of the most important base monomers for the production of acrylate adhesives [13].

2.8 Hydroxyethyl methacrylate

Hydroxyethyl methacrylate (HEMA), also named as 2-hydroxyethyl methacrylate, 2-hydroxyethyl-2-methylprop-2-enoate, glycol methacrylate, glycol monomethacrylate, 2-(methacryloyloxy)ethanol or ethylene glycol methacrylate, presents the molecular formula $(C_6H_{10}O_3)_n$.

HEMA is a colorless and viscous liquid, which can readily polymerize and used as monomer for the production of various types of polymers. It is hydrophobic in nature, nevertheless, in the form of hydrogel, if exposed to water, it swells as a result of hydrophilic pendant group. HEMA is capable of absorbing from 10 to 600% of water relative to its dry weight depending on its chemical and physical structure. For this reason, it is usually used in the production of soft contact lenses [13].

When treated with polyisocyanates, HEMA makes a crosslinked polymer, an acrylic resin that is a useful component in some paints. HEMA can also be used as a matrix with 40 nm silica particles for 3D glass printing. It is one of the several polymeric hydrogels that are biocompatible with adjustable mechanical properties including resistance to crack propagation [20]. As the first hydrogel ever used for biomedical application [20, 21], it has been utilized in the production of stents and catheters [21]. Its porous structure eases the diffusion of fluids including oxygen making it suitable for fabrication of hydrophilic contact lens and artificial cornea prostheses [20].

2.9 Poly(2-hydroxyethyl methacrylate)

Poly(2-hydroxyethyl methacrylate) (PHEMA) is one of the major biomaterial components of most soft contact lenses and is also part of intraocular lenses because of their inherent optical properties [21]. This is due to their hydroxyl groups (–OH groups) being free. PHEMA hydrogels contains very high amounts of water, also posing hydrophilic properties [22, 23], facilitating the diffusion of oxygen and solutes. That is why oxygen can gain access to the eyeballs through its gel aqueous phase without any hindrance in contact lenses. Lenses that are produced with PHEMA as one of the major constituting materials. Thus, they are widely used because the polymer has a unique blend of good stability to varying conditions of pH and mechanical properties, ionic strength and temperature [21]. Considering its suitability for ocular lenses, PHEMA has exhibited excellent biocompatibility. Hydrogels made from PHEMA and its co-polymers have been critically employed for biomedical uses and considered for controlled drug delivery applications. PHEMA gels have been used in several attempts to reconstruct nasal cartilages, female breasts, as artificial corneas and as wound dressings.

2.10 Butyl acrylate

Butyl acrylate (BA), also called as butyl prop-2-enoate, n-butyl acrylate, butyl ester of acrylic acid or butyl-2-propenoate, presents the molecular formula $(C_7H_{12}O_2)_n$.

BA can be synthesized in various different reactions. Thus, 1-butyl alcohol, acetylene, carbon monoxide, hydrochloric acid and nickel carbonyl can react to yield butyl acrylate. Other chemical route in the synthesis of BA comprises the reaction of methyl acrylate with butanol or acrylic acid.

Butyl acrylate is used in coatings, paints, adhesives, sealants, plastics, textiles, caulking materials and fuel [24]. Its derivatives could be used to improve the impact strength of other acrylic plastics, by incorporating them as impact modifiers [25].

2.11 Trimethylolpropane Triacrylate

Trimethylolpropane Triacrylate (TMPTA), also named as 2,2-Bis[(acryloyloxy) methyl]butyl acrylate, presents the molecular formula (C15H20O6)n.

TMPTA is a tri-functional monomer, which is used in the production of adhesives, plastics, inks, anaerobic sealants and acrylic glue. Its usefulness stern from its low volatility and fast cure rate. It possesses good water, chemical, weather and abrasion resistance. The end products of TMPTA include hardwood floors, alkyd coatings, concrete polymers, compact discs, dental polymers, letterpress, lithography, automobile headlamps, elastomers, acrylics, screen printing and different plastic components for medical industry [3, 4, 6].

2.12 Sodium polyacrylate

Sodium polyacrylate (NaPA), also named as poly (sodium prop-2-enoate), presents the molecular formula $(C_3H_3NaO_2)_n$. NaPA is a sodium salt of polyacrylic acid with broad application in consumer products, and structural engineering [26]. Though it is soluble in water, when processed into a chemically crosslinked polymer hydrogel, it would swell greatly in water [27]. This occurs as a result of the sodium carboxylate salt present in water and the interactions of the hydrogen bonds between water and the crosslinked polymer. This super osmotic process of water absorption tagged this material as a super absorbent (also known as water lock) polymer (SAP) because it is capable of absorbing up to 100–1000 times its weight when exposed to water. When this polymer is blended with clay, it renders improved and controlled swelling properties [28].

Sodium polyacrylate is an anionic polyelectrolyte-based polymer comprising of carboxylic groups that are negatively charged in the main chain. There are other salts neutralized polyacrylic acids available in the market including potassium, lithium and ammonium. However, sodium neutralized polyacrylic acids are the most common form used in industry.

NaPA and other derivatives of polyacrylic acid have varied commercial and industrial applications: animal care (pet pads, horse urine odor absorbing materials, drown-free water source for feeder insects); topical health (paper and disposable diapers, sanitary wares and napkins); industry (waste liquid control, drilling fluids, concrete protections). environmental protection (geo-textiles for road constructions, anti-flood bags, excreta collection bags and containers); structural engineering application of concrete structures with sodium polyacrylate serving as an internal curing agent and exhibiting enhanced durability and strength performance [26]; biomedical field for drug delivery applications in ocular, nasal, buccal, gastrointestinal, epidermal and transdermal drug delivery system due to their biocompatibility and similarities with living tissues in the pharmaceutical industry [27]; other products such as wire and cables, water blocking, artificial snows, hot/cold gel packs, urine bags, toys, thickening agents, fragrance carriers, fire-retardant gels, anti-fogging packing materials and waterbeds.

2.13 Poly(N-isopropylacrylamide)

Poly(N-isopropylacrylamide) (PNiPAAm) has gained significance as injectable polymer in cell and drug delivery due to the unique physicochemical properties. PNiPAAm undergoes phase separation resulting in the formation of an opaque hydrogel in response to a temperature above 32°C. This thermoresponsive behavior is as a result of strong hydrogen bonds between water molecules, the polymer and the specific molecular orientations of the bonds due to the molecular structure of this polymer.

2.14 N-butyl acrylate

N-Butyl Acrylate (BA) is a soft monomer with low T_g and regarded as the largest-volume acrylate ester used in the preparation of all styrene acrylic copolymers, acrylic and vinyl acrylic. The other major acrylate esters include 2- ethylhexyl acrylate (2-EHA), ethyl acrylate (EA), and methyl acrylate (MA). Butyl acrylate is used as a soft-monomer to improve toughness and low temperature. The major applications for BA are in coatings and paints within the automotive and architectural field, sealants, adhesives and intermediates. Other areas of BA application include leather finishes, inks, caulks, textiles and papers.

Another important and growing usefulness of BA is in thermoplastic ethylene acrylate co-polymers (EAC). They are used as processing aid and impact modifier in thermoplastics to improve properties such as flexibility, toughness, molding characteristics, part appearances and low temperature properties. End use applications include adhesives, packaging and multilayer films. BA showcases superior photostability and is a preferred monomer where weatherability, with emphasis on low temperature applications especially at high altitudes, and sunlight resistance are required [24, 29, 30].

The combination of butyl acrylate with other polymerizable acrylate ester monomers allows the fabrication of copolymer compositions. This way, the performance of co-polymers can be tailored to meet a wide range of end-use requirements as it has the capacity to balance softness and hardness, block resistance and tackiness, low temperature flexibility, durability and strength, and other important properties required of products to meet end-use goals.

3. Possible risks associated with the use of acrylates

The U.S. Environmental Protection Agency (US EPA) as well as the International Agency of Research on Cancer have classified some acrylates as a possible human carcinogen [11, 31]. For example, in dentistry, the use of acrylates as fillings may cause dental complications like pulpitis and periodontitis as a result of heat generation in the course of curing, as well as acidic and porous conditions [32]. Other exposures to acrylates have been related to eye, throat and skin reactions as well as more serious health consequences such as reproductive toxicity, cancer, neurological damage, development issues, organ system toxicity, cellular damage. Nonetheless, several acrylate polymers are currently successfully used or under research in a wide range of biomedical applications such as contact lenses, bone cements and scaffolds for tissue engineering [33–41].

4. Conclusion

Acrylic polymers and co-polymers are currently widely used in industry in all kind of applications due to their chemical purity, stability, high heat resistance, sunlight resistance, excellent weathering, low temperature performance, water resistance and hydrophobicity. Acrylates have found application in the manufacture of co-polymers for coatings and paints, sealants, adhesives, textile fibers, printing inks. It is highly superabsorbent polymers and thus used in diapers. Thermoplastic acrylate co-polymers, in biomedicine and a variety of other advanced application areas.

Author details

Kingsley Kema Ajekwene
Department of Polymer and Textile Technology, Yaba College of Technology, Yaba, Lagos, Nigeria

*Address all correspondence to: kemaking@gmail.com

References

[1] Brandrup J, Immergut EH, Grulke EA, editors. Polymer Handbook. 4th ed. New York: John Wiley & Sons; 1999

[2] Fenichell S. Plastic: The Making of a Synthetic Century. New York: Harper Collins; 1996

[3] Seymour RB, Carraher CE. Giant Molecules. New York: John Wiley and Sons, Inc.; 1990

[4] Ulrich H. Introduction to Industrial Polymers. Munich: Hanser; 1982

[5] Ohara T, Sato T, Shimizu N, Schwind GPH, Weiberg O, Marten K, et al. Acrylic Acid and Derivatives. Ullmann's Encyclopedia of Industrial Chemistry. Weinheim: Wiley-VCH; 2003. DOI: 10.1002/14356007.a01-161

[6] Ervithayasuporn V, Chimjarn S. Synthesis and isolation of methacrylate- and acrylate-functionalized polyhedral oligomeric silsesquioxanes (T8, T10, and T12) and characterization of the relationship between their chemical structures and physical properties. Inorganic Chemistry. 2013;**52**(22):13108-13112. DOI: 10.1021/ic401994m

[7] Stickler M, Rhein T. Polymethacrylates. In: Ullmann's Encyclopedia of Industrial Chemistry. Weinheim: Wiley-VCH; 2000. DOI: 10.1002/14356007.a21-473

[8] The Royal Society of Chemistry. Chemistryworld. Available from: http://www.rsc.org/chemistryworld [Accessed: 05 March 2019]

[9] Bhanu VA, Rangarajan P, Wiles KB, Bortner MJ, Sankarpandian M, Godshall D, et al. Synthesis and characterization of acrylonitrile methyl acrylate statistical copolymers as melt processable carbon fiber precursors. Polymer;**43**(18):4841-4850. DOI: 10.1016/S0032-3861(02)00330-0

[10] Rey-Vinolas S, Engel E, Mateos-Timoneda MA. Polymers for bone repair. In: Bone Repair Biomaterials. Woodhead Publishing Series in Biomaterials. UK: an imprint of Elsevier; 2019. pp. 179-197. DOI: 10.1016/ b978-0-08-102451-5.00007-x

[11] Chalifoux PR. Acrylic and other resins: Provisional restorations. In: Esthetic Dentistry. 2015. pp. 197-230. DOI: 10.1016/b978-0-323-09176-3.00019-x

[12] da Silva Júnior WF, de Oliveira Pinheiro JG, Moreira CDLFA, de Souza FJJ, de Lima ÁAN. Alternative technologies to improve solubility and stability of poorly water-soluble drugs. In: Multifunctional Systems for Combined Delivery, Biosensing and Diagnostics. UK: Elsevier BV; 2017. pp. 281-305

[13] Dow. Technical Data Sheet. Ethyl Acrylate. Available from: www.dow.com [Accessed: 05 March 2019]

[14] Hacker MC, Krieghoff J, Mikos AG. Synthetic polymers. In: Principles of Regenerative Medicine. UK: Elsevier BV; 2019. pp. 559-590. DOI: 10.1016/b978-0-12-809880-6.00033-3

[15] Campaign for Safe Cosmetics. A Project of Breast Cancer Prevention Partners. Available from: http://www.safecosmetics.org/get-the-facts/chemicals-of-concern/2978/ [Accessed: 05 March 2019]

[16] Steven GV, Stott DE. Polyacrylamide: A review of the use, effectiveness, and cost of a soil erosion control amendment. In: Sustaining the Global Farm. Selected Papers from the

10th International Soil Conservation Organization Meeting; 2001

[17] IGTPAN. Instituto Granado de Tecnología de Poliacrilonitrila. Properties of Polyacrylonitrill. Available from: http://www.igtpan.com/Ingles/propriedade-poli.asp [Accessed: 15 April 2019]

[18] IGTPAN. Instituto Granado de Tecnología de Poliacrilonitrila. Polyacrylonitrile Applications. Available from: http://www.igtpan.com/Ingles/aplicacoes.asp [Accessed: 15 April 2019]

[19] Gupta BS, Afshari M. Handbook of properties of textile and technical fibres. The Textile Institute Book Series. 2nd ed. UK: Woodhead Publishing Ltd; 2018. pp. 545-593. DOI: 10.1016/B978-0-08-101272-7.00015-8

[20] Ananya B. Synthetic polymeric gel. In: Polymeric Gels. UK: Woodhead Publishing Ltd; 2018. pp. 55-90. DOI: 10.1016/b978-0-08-102179-8.00003-x

[21] Dupont. DuPontTM Elvaloy® AC Products and Properties. Available from: https://www.dupont.com [Accessed: 15 April 2019]

[22] Bartkowiakb G, Frydrych I. Superabsorbents and their medical applications. In: Handbook of Medical Textiles. UK: Woodhead Publishing Ltd; 2011. pp. 505-544

[23] Malmonge SM. Biocompatible hydrogels. In: Reference Module in Materials Science and Materials Engineering. UK: Elsevier Ltd; 2018. DOI: 10.1016/b978-0-12-803581-8.10077-3

[24] Stoye D, Funke W, Hoppe L. Paints and Coatings, Ullmann's Encyclopedia of Industrial Chemistry. Weinheim: Wiley-VCH; 2006. DOI: 10.1002/14356007.a18-359

[25] Bortnick NM. Polyacrylics. In: Encyclopedia of Materials: Science and Technology. UK: Elsevier Ltd; 2001. pp. 7144-7147. DOI: 10.1016/b0-08-043152-6/01265-1

[26] Manzur T, Iffat S, Noor MA. Efficiency of sodium polyacrylate to improve durability of concrete under adverse curing condition. Advances in Materials Science and Engineering. 2015;**2015**:1-8. DOI: 10.1155/2015/685785

[27] Sultana S et al. Swelling and physico-mechanical properties of synthesized sodium polyacrylate hydrogels. International Journal of Advanced Research. 2017;**5**(7):84-92. DOI: 10.21474/ijar01/4696

[28] Takeno H et al. Mechanical, swelling, and structural properties of mechanically tough clay-sodium polyacrylate blend hydrogels. Gels. 2017;**3**(1):10. DOI: 10.3390/gels3010010

[29] Gantrade. Applications of n-Butyl-Acrylate. Available from: https://www.gantrade.com [Accessed: 05 March 2019]

[30] Zondlo Fiume M. Final report on the safety assessment of acrylates copolymer and 33 related cosmetic ingredients. International Journal of Toxicology. 2002;**21**(3):1-50. DOI: 10.1080/10915810290169800

[31] World Health Organization: IARC. Monographs on the Evaluation of the Carcinogenic Risk of Chemicals to Humans. Geneva: World Health Organization: IARC; 1999

[32] Serrano-Aroca Á, Llorens-Gámez M. Dynamic mechanical analysis and water vapour sorption of highly porous poly(methyl methacrylate). Polymer. 2017;**125**:58-65

[33] Serrano-Aroca Á, Campillo-Fernández AJ, Gómez-Ribelles JL, Monleón-Pradas M, Gallego-Ferrer G, Pissis P. Porous poly(2-hydroxyethyl acrylate) hydrogels prepared by radical

polymerisation with methanol as diluent. Polymer. 2004;**45**:8949-8955

[34] Monleón-PradasM,Gómez-RibellesJL, Serrano-Aroca A, Ferrer-Gallego G, Anton JS, Pissis P. Interaction between water and ploymer chains in poly (hydroxyethyl acrylate) hydrogels. Colloid & Polymer Science. 2001;**279**:323-330

[35] Monleón-Pradas M, Gómez-Ribelles JL, Serrano-Aroca A, Gallego-Ferrer G, Suay-Antón J, Pissis P. Interaction between water and polymer chains in poly(hydroxyethyl acrylate) hydrogels. Colloid and Polymer Science. 2001;**279**:323-330

[36] Serrano-ArocaÁ,Pradas-MonleónM, Ribelles-Gómez JL. Macroporous poly(methyl methacrylate) produced by phase separation during polymerisation in solution. Colloid and Polymer Science. 2007;**285**(7):753-760

[37] Serrano-Aroca A, Pradas MM, Ribelles JLG, Rault J. Thermal analysis of water in reinforced plasma-polymerised poly (2-hydroxyethyl acrylate) hydrogels. European Polymer Journal. 2015;**72**:523-534

[38] Brígido-Diego R, Pérez-Olmedilla M, Serrano-Aroca A, Gómez-Ribelles JL, Pradas-Monleón M, Gallego-Ferrer G, et al. Acrylic scaffolds with interconnected spherical pores and controlled hydrophilicity for tissue engineering. Journal of Materials Science. Materials in Medicine. 2005;**16**:693-698

[39] Serrano-Aroca Á, Monleón-Pradas M, Gómez-Ribelles JL. Plasma-induced polymerisation of hydrophilic coatings onto macroporous hydrophobic scaffolds. Polymer. 2007;**48**:2071-2078

[40] Serrano-Aroca Á, Monleón-Pradas M, Gómez-Ribelles JL, Vidaurre-Garayo A, Suay-Antón J. Characterisation of macroporous poly(methyl methacrylate) with plasma-polymerised poly(2-hydroxyethyl acrylate) coating. European Polymer Journal. 2007;**43**:4552-4564

[41] Serrano-Aroca A, Monleón Pradas M, Gómez Ribelles JL. Effect of crosslinking on porous poly(methyl methacrylate) produced by phase separation. Colloid and Polymer Science. 2008;**286**:209-216. DOI: 10.1007/s00396-007-1755-0

Section 2

Acrylate Polymers in Biomedicine

Chapter 4

Acrylic-Based Materials for Biomedical and Bioengineering Applications

Ángel Serrano-Aroca and Sanjukta Deb

Abstract

Acrylic-based polymers have been used for many years in biomedical applications because of their versatile properties. Many different polymers belong to this class of polymers, of which a significant number have been approved by the US Food and Drug Administration (FDA) and are frequently used in ophthalmologic devices, orthopaedics, tissue engineering applications and dental applications. The applications of this class of polymers have the potential to be expanded exponentially in the biomedical industry if their properties such as mechanical performance, electrical and/or thermal properties, fluid diffusion, biological behaviour, antimicrobial capacity and porosity can be tailored to specific requirements. Thus, acrylic-based materials have been produced as multicomponent polymeric platforms as interpenetrating polymer networks or in combination with other sophisticated materials such as fibres, nanofibres, carbon nanomaterials such as graphene and its derivatives and/or many other types of nanoparticles in the form of composite or nanocomposite biomaterials. Moreover, in regenerative medicine, acrylic porous supports (scaffolds) need to be structured with the necessary degree, type and morphology of pores by advanced technological fabrication techniques.

Keywords: acrylic-based materials, biomedicine, bioengineering, composites, nanocomposites, tissue engineering

1. Introduction

Acrylic-based materials have been used in medicine and dentistry over decades. Specific examples in current clinical use are corneal prosthesis, intraocular lenses and contact lenses in ophthalmology [1] and bone cements for orthopaedic applications [2], tissue engineering [3] and dental restoratives [4] because of their versatile properties [5]. A significant number of these acrylic polymers have been affirmed by the US Food and Drug Administration (FDA) for different applications. Nonetheless, a considerable number of potential clinical applications are impeded by their inadequate mechanical properties, biological behaviour, electrical and/or thermal capacity, interaction with body fluids, antimicrobial response and porosity in porous platforms (scaffolds) for tissue engineering applications. To address some of these limitations of existing acrylates, new acrylic-based materials with superior properties have been produced and are currently under being researched to resolve these problems by means of diverse fabrication strategies such as multicomponent polymeric systems or by incorporation of other materials and/or nanomaterials such as graphene to create composites or nanocomposites with a wide range of properties and also develop extensive interconnected porous structures. Biomaterials are used to engineer functional restoration of different tissues towards the repair or restoration of damaged tissues via interacting with living systems and enabling normal function. Hence the design and development of novel biomaterials involve considering physical, mechanical and biological properties of the native tissue that is being replaced with a goal of mimicking it as closely as possible to enable repair or replace and function. One of the important aspects is designing the material with desired mechanical properties that enables survival in a complex biological milieu. Recent technological advances have seen the development of complex biomaterials, and newer manufacturing techniques allow for increased spatial control over material properties. Despite the current advances, the improvement of mechanical properties of acrylics still remains a challenge and needs further investigation [6–11].

2. Mechanical performance

The enhancement of the mechanical performance of acrylics is one of the hot themes in the area of biomedical engineering, and numerous researchers have been working on this field for many years. Acrylics can be reinforced through different methods and techniques: by incorporating ceramic additives to form composites [12]; microphase-separated morphologies such as block copolymers, in which hydrophobic and hydrophilic domains alternate [13]; increasing the cross-linking density [14]; binary systems composed of two or more mixed polymers as interpenetrating polymer networks [15, 16]; reinforced acrylic-based composites with embedded fibres [17]; plasma-induced polymerisation of an acrylic hydrogel onto a hydrophobic acrylic support [7, 18]; and nanosilica reinforcement obtained through sol-gel reactions [19]. Nevertheless, new procedures to enhance the mechanical behaviour of acrylics have been performed by incorporating advanced materials such as graphene (2010 Nobel Prize in Physics) [20] and other outstanding carbon-based materials such as carbon nanotubes (CNT) [21]. Graphene derivatives such as graphene oxide (GO) [22–24] or reduced GO (rGO) [25] have also exhibited excellent reinforcement for acrylics, especially hydrated acrylic hydrogels, and enhancement of many other properties.

2.1 Interpenetrating polymer networks

Interpenetrating polymer networks (IPN) as reinforced polymer networks containing acrylic polymers have gained much attention in the biomedical field during the last few decades. An IPN consists of two separate but interwoven polymer networks, which can enhance and/or combine functional properties. In this advanced multicomponent polymeric system, there are not any covalent bonds between them, but at least one of them is cross-linked within the immediate presence of the other. In this field, there are six basic multicomponent polymeric structures: mechanical blends, block copolymers, graft copolymers, AB-cross-linked copolymer, semi-IPNs and full IPNs [26]. A full IPN results is produced if a cross-linker is present in the polymeric system [27]. However, a semi- or pseudo-IPN polymer network possessing linear polymers embedded within the first cross-linked network is formed in the absence of cross-linking [28, 29]. Simultaneous interpenetrating polymer networks (SINs) are produced when the precursors of both networks are mixed and the two networks are formed at the same time or in the form of sequential IPNs, by swelling of a single-polymer network into a solution containing the mixture of monomer, initiator, activator and cross-linker. Thus, for example, urethane acrylate resin networks have been reinforced with epoxy networks in the form of SINs [30]. Full IPNs and semi-IPNs have been prepared with epoxy resin and poly(ethyl methacrylate) (PEMA) utilising the sequential mode of synthesis [31]. Both types of networks showed a gradual decrease of modulus and tensile strength performance and an increase of elongation at break and toughness with increasing PEMA content. Simultaneous semi-interpenetrating polymer networks (semi-SINs) of epoxy resin-acrylate polyurethane with high compatibility have been synthesised by simultaneous photopolymerisation [32]. More recently oxirane-acrylate IPN's were reported to enhance the hydrophobic nature of dental composites and lower degradation and shrinkage stress than commonly used bis-GMA resins; however, the mechanical properties were not adequate [16]. Pseudo-SIPNs have been prepared by melt blending of poly(methyl methacrylate) (PMMA) and double-C_{60}-end-capped poly(ethylene oxide) (FPEOF) that exhibited a storage modulus 16 times higher than PMMA, which are as similar to nanocomposites of PMMA with carbon nanotubes [29].

Acrylic-based IPN hydrogels are also produced with the goal of improving the mechanical performance and swelling-deswelling action of acrylic hydrogels [33]. In this way, for instance, hydrophilic interpenetrating polymer networks (IPN) with good mechanical properties even in the hydrated state were produced with chitosan and poly(acrylic acid) (PAA) [34].

2.2 Composite biomaterials

Most chemical modifications of acrylic hydrogels do not result in a significant change of the overall mechanical properties since the main structural skeletons of these polymers or copolymers remain weak. On the contrary, the reinforcement achieved by the incorporation of fibre into acrylic polymer matrices is different because the added fibres impart high strength to the networks with the main skeleton of the composites. Materials reinforced with fibre are composite materials consisting of a polymer phase embedded with high-strength fibres, such as glass, aramid and carbon [35]. In this type of materials, the mechanical performance is significantly enhanced, and the biocompatible properties of the acrylic polymer phase should remain unmodified. Thus, acrylic resin polymers have been reinforced with glass fibres for dental applications [36], and acrylic hydrogels such as poly(2-hydroxyethyl methacrylate), which is one of the most popular biomaterials, have been used to manufacture fibre-reinforced composites by inclusion of various kinds of woven and knitted fabrics and fibres, in order to improve overall qualities of the PHEMA-based artificial skin for wound dressing applications [37]. Nevertheless, in the last recent decades, natural fibres have attracted much attention as reinforcement agents for polymer-based composites due to their advantages over other types of conventional fibres made with glass or carbon [38]. Thus, natural fibres, such as flax, hemp, sisal, jute, kenaf, coir, kapok and banana, among many others, have shown many advantages over man-made glass and carbon fibres: lower cost, lower density, comparable specific tensile performance, less energy consumption, non-abrasive, not irritating, lower health risk, renewable, biodegradable and recyclable [39]. Thus, in the field of acrylic polymers, ultra-long chitin natural fibres obtained in a sustainable way, that is, from marine-river crab shell wastes, have been incorporated into PMMA resins to produce nanocomposites with outstanding properties for biomedical and bioengineering applications [40].

2.3 Nanocomposite biomaterials

Nanocomposites are of particularly interest due to the fact that they exhibit both individual properties of the polymer and properties due to the presence of nanoparticles, and this additive effect has led to several therapeutic and diagnostic applications [41]. Another reinforcing alternative for acrylic polymers consists of adding nanomaterials such as silica, carbon nanomaterials, nanofibres or other nanoparticles. Silica is biocompatible, possesses bioactive properties [42] and can improve the mechanical behaviour of acrylics through filling or by the sol–gel process [43]. Thus, for example, biphasic matrices of hybrid nanocomposite materials consisting of an organic phase of poly(2-hydroxyethyl acrylate) and an inorganic phase of silica network obtained by the sol–gel process of tetraethoxysilane (TEOS) exhibited a very significant enhancement of the mechanical properties [9]. The combination of interpenetrated polymer networks and nanosilica filling has been explored, and simultaneous polyurethane/PEMA interpenetrating polymer networks with silica filler consisting of very fine powders with an approximate diameter of 5 nm were reported to significantly improve material strength [44]. Polymer nanocomposites have significantly influenced the field of biomedicine. In this regard, graphene-based nanocomposites have been reported to enhance the physical and biological properties of a wide range of polymers of different chemical natures [24, 45–53]. Graphene (GN) is a 2D monolayer of sp^2-bonded carbon atoms [54] very promising in a wide range of industrial fields due to its outstanding electrical and thermal properties [55, 56] and high mechanical performance [57]. In addition, it has been demonstrated that graphene, in the biomedical field, promotes adherence of human osteoblasts and mesenchymal stromal cells [58]. Graphene has exhibit potential application in the development of new reinforced nanocomposites such as magnetite-GNs/poly(arylene-ether-nitrile) [59]. Furthermore, GN is able to enhance the shape memory of acrylic polymers such as poly(acrylamide-co-acrylic acid), and it possesses self-healing ability when its content is between 10 and 30% [60]. The oxidised form of GN, graphene oxide (GO), can also improve the mechanical strength and modulus of acrylic polymer networks such as poly(acrylamide) (PAM) hydrogels [61]. GO is also a two-dimensional nanomaterial usually produced from natural graphite by exfoliation into monolayer sheets. GO possesses oxygenated functional groups at the basal plane and at the edges, hydroxyl (–OH), epoxy (–C–O–C–), carboxyl (–COOH) and carbonyl (–C=O), which render its dispersion in aqueous solution easier [62]. The outstanding unique properties of GO, such as high tensile modulus (1.0 TPa), ultimate strength (130 GPa) and electrical and thermal properties [63], render this carbon nanomaterial an ideal candidate for a broad range of applications in the field of new advanced materials. Thus, for instance, when GO nanosheets are introduced into weak and brittle PAM hydrogels, their mechanical properties improve significantly [61]. The incorporation of GO nanosheets into poly (acrylic acid)/gelatin composite hydrogels increased significantly their Young's modulus and maximum stress. In this study, the hydrogel with GO (0.2% *w/w*)/PAA(20% *w/w*) exhibited the highest Young's modulus, whereas GO (0.2% *w/w*)/PAA (40% *w/w*) composites showed the highest maximum stress. These results suggest that GO nanosheets can be successfully used as reinforcing agents to enhance the mechanical performance of hydrogel materials, which is often required for certain tissue engineering applications [23]. Other graphene derivatives such as chemically modified graphene (CMG) have been proposed and compared with GO for the reinforcement of PMMA [64]. Single-wall carbon nanotubes (SWCNTs), multiwalled carbon nanotubes (MWCNTs) and carbon nanofibres (CNFs) are also very promising reinforcing nanomaterials for acrylic polymers such as poly(methyl methacrylate). For instance, nanocomposites of PMMA with small amounts of

CNFs (5% *w/w*) can enhance over 50% of the axial tensile modulus with respect to neat PMMA. PMMA/CNF nanocomposites in the form of fibres have also exhibited enhanced thermal stability, reduced shrinkage and improved modulus retention with temperature, as well as stronger compression performance [65]. The electrical and electromechanical properties of acrylic materials such as acrylic elastomers and styrene copolymers can also be enhanced to be suitable for electroactive applications such as artificial muscles and/or microelectromechanical system (MEMS) devices [66]. Natural chitosan nanofibres can act as a multifunctional cross-linker and a reinforcing agent in poly(acrylamide) (PAM) nanocomposites, which were prepared via in situ free-radical polymerisation, producing superior compression strength and storage modulus than those of pure PAM [67]. Plant fibre-based nanofibres using fibrillation of wood pulp fibres into nanofiber bundles, which are thin enough to work, as well as bacterial cellulose in maintaining the transparency of resin also enhanced the physical properties of the composites [68]. Nanoparticles such as clay can also be used to enhance the mechanical properties of acrylic polymers. Thus, for instance, reinforced PMMA/clay nanocomposites exhibit a polymer matrix reinforced by usually 10% *w/w* or less of uniformly dispersed inorganic particles with at least one dimension in the nanometre scale [69]. These nanocomposites have also shown enhanced thermal properties when compared to pure polymer or common composites. Biocomposites based on acrylated fatty acids and magnetic nanoparticles have been reported to maintain their good magnetic response and supramagnetic behaviour, and the use of acrylated fatty acids further enables chemical modification of magnetic nanoparticles extending their potential applications [70]. The modification of graphene with biomacromolecules like DNA, protein, peptide and others extends the potential applications of graphene materials. The remarkable advances in material technologies and recent developments in nanotechnology have enabled formulation of different nanocomposites with a wide range of properties, and the potential of application of these composites in healthcare is becoming more apparent. Nanocomposites are already being successfully used in drug delivery, gene therapy, tissue engineering and imaging and as biosensors [70–74].

3. Electrical properties

The electrical properties of acrylics are important in some biomedical and bioengineering fields because various types of electrical stimulation can regulate cell physiological activities such as division [75], migration [76], differentiation and cell death [77]. Promotion of spinal cord repair and successful cancer therapy has been also achieved by a combination of electrical stimulation and the non-invasive nature of these polymers [78–80]. For that reason, the development of new conductive acrylic-based materials for biomedical applications is gaining much attention during the last decades. Graphene, a nanocarbon material, has been recently very effective as an electrode material with very high conductivity [41]. Although graphene exhibits high transmittance and excellent conductivity [55], its fabrication is still expensive. However, GO, which is more cost-effective, possesses low conductivity due to its oxygen-containing functional groups, and it must be modified to obtain reduced graphene oxide (rGO) in order to develop electrically conductive acrylic-based resins. Thus, for instance, single-step procedures can be performed starting from a homogeneous water dispersion of GO in order to effect a reduction induced by the UV radiation during the photopolymerisation of an acrylic resin [81]. Transparent conductive films can be also fabricated by grafting of poly(acryl amide)/poly(acrylic acid) onto the GO surface followed by a reduction to rGO nanosheets by a two-step chemical reduction [82]. Flexible and conductive double networks (DN) of rGO and poly(acrylic acid) hydrogels have also been synthesised by a two-step preparation with a reduction-induced in situ self-assembly [83]. Nacre-inspired acrylic-based nanocomposites of rGO and PAA have also been prepared via a vacuum-assisted filtration self-assembly process. Higher strength and toughness (2 and 3.3 times higher) were achieved with these bioinspired nanocomposites in comparison with those of the pure reduced GO film due to the high amount of hydrogen bonds between the GO nanosheets and PAA polymer chains. Furthermore, this nanocomposite also displayed high electrical conductivity (108.9 $S{\cdot}cm^{-1}$), which renders it a promising biomaterial for many advanced applications in biomedicine and bioengineering, such as flexible electrodes and artificial muscles. The one-dimensional carbon nanomaterials discovered by Iijima [84], named as carbon nanotubes (CNTs), have caused increasing interest due to their excellent electrical conductivity and remarkable mechanical properties with numerous potential nanotechnological applications [85]. Ultrasensitive electrochemical biosensors can be developed with CNTs due to their unique electrical properties. Thus, glucose biosensor for diabetics has been developed with nanofibrous membranes of poly(acrylonitrile-co-acrylic acid) (PANCAA) filled with multiwalled carbon nanotubes (MWCNT) [86]. Other acrylic polymers such as poly(methyl methacrylate) have been developed as nanocomposites containing various multiwalled carbon nanotube (MWCNT) contents by melt mixing to achieve high conductivity levels [87].

4. Thermal properties

The enhancement of the thermal properties of acrylics can increase its long-term operation at the human body temperature. Thus, for instance, semi-IPNs based on polyurethane and poly(acrylamide) exhibited enhanced thermal properties due to higher cross-link density produced by the hard segment content [28]. Differential scanning calorimetry PHEMA/SiO_2 hybrids are complicated showing complex results with two glass transition temperatures (*Tg*). In addition, the SiO_2 content showed to be an important factor in influencing the *Tg* shift of the thermal transition [88]. Nevertheless, polymer nanocomposites with functionalized graphene sheets (FGNS) exhibited an unprecedented shift in *Tg* of up to 40°C and 30°C in poly(acrylonitrile) with 1% *w/w* of FGNS and in PMMA with only 0.05% *w/w*, respectively [89]. Furthermore, the thermal stability of magnetite-graphene/poly(arylene-ether-nitrile) nanocomposites was significantly improved by the addition of magnetite-graphene hybrids [59]. PMMA-based nanocomposites produced with CMG and GO by in situ polymerisation also exhibited large shifts in the *Tg* with loadings as low as 0.05% *w/w* [64]. Another approach to enhance the thermal behaviour of acrylic polymers is by incorporating nanoparticulate fillers. Thus, the thermal performance of common acrylic polymers such as PMMA has been significantly enhanced by filling with several percentages (5%, 10%, 15% and 20%) of nanometric particles of titanium oxide (TiO_2) and ferric oxide (Fe_2O_3) by the solvent casting method [90, 91]. Thermal degradation can also be enhanced in acrylic polymers by the addition of nanoparticulate fillers. Thus, the thermogravimetric analysis (TGA) of PMMA-TiO_2 and PMMA-Fe_2O_3 composites showed that these nanoparticles can improve the thermal stability of neat PMMA by about 50°C with just a low filling of 5% *w/w* [91]. TGA has also shown that the presence of very low amounts of Pd nanoparticles (0.0005–0.005 vol%) in PMMA can significantly enhance its thermal stability, as shown by a 75°C degradation initiation retardation and a 32°C gain at the maximum decomposition rate [92]. Acrylic hydrogels as hydrophilic polymers are able to absorb large amounts of water in biomedicine due to their contact with fluids in cells or tissue in the human body. Therefore, the thermal analysis of water and its influence on the hydrated hydrogel performance becomes essential. Thus, many studies have been reported in this way with acrylic hydrogels such as PHEMA [93], bulk and plasma-poly(2-hydroxyethyl acrylate) (PHEA) [6] and poly(ethyl acrylate) [94].

5. Water sorption and diffusion

Water sorption and diffusion are also very important in biomedicine because these properties play a very important role in cell survival, dimensional changes and release of small molecules which are all factors that need to be considered especially in tissue engineering applications [3]. Thus, acrylic hydrogels such as poly(2-hydroxyethyl methacrylate) or poly(2-hydroxyethyl acrylate) are important due to their hydrophilicity, swelling and deswelling capacity [95–97]. Their outstanding water sorption properties have made this kind of hydrophilic materials very promising for a broad range of biomedical and bioengineering applications such as wound healing, controlled drug delivery, tissue engineering, etc. [5, 98]. The hydrophilic functional groups attached to the polymeric backbone of these polymers provide the ability to absorb water, whilst their resistance to dissolution arises from cross-linking of polymer chains [99]. Nevertheless, these single-network hydrogels possess very weak mechanical performance and slow swelling response. Therefore, reinforcement strategies, such as the combination of hydrophilic and hydrophobic functional groups as multicomponent polymeric systems, are usually required for these types of polymer, which can also decrease their water sorption. The reinforcement of acrylics through the incorporation of GO nanosheets can also modify their water sorption behaviour. Thus, for example, the swelling rates of GO/poly(acrylic acid-co-acrylamide) nanocomposite hydrogels increased with increasing GO content to about 0.30% *w/w* and then decreased with further increase in amounts of GO. It is worth noting that the hydrogel with only 0.10% *w/w* GO exhibited significant improvement in swelling capacity in a neutral medium and could retain relatively higher swelling rates in acidic and basic solutions. Therefore, these GO-based superabsorbent acrylic hydrogels have potential applications in biomedical engineering and hygiene products [62]. The mechanism of water diffusion [100] can also be altered by reinforcement of acrylics through any of the methods discussed in Section 1. Thus, poly(acrylic acid)-GO nanocomposite hydrogels, which are potential carriers for drug release, can be manipulated by changing the concentration of GO and tend to exhibit non-Fickian anomalous diffusion with decrease in deswelling ratio with increasing GO content [63]. There are many acrylic hydrogels, which exhibit non-Fickian diffusion behaviour such as poly(2-hydroxyethyl acrylate) [97, 101]. Important water-swellable biomedical polymer such as PHEMA has shown to be governed by Fickian diffusion, even though water sorption is not classically Fickian [102]. Thus, advanced hydrogels based on 2-hydroxyethyl methacrylate (HEMA) and epoxy methacrylate (EMA) produced via bulk polymerisation exhibited also a Fickian swelling process, and the equilibrium water content (EWC) decreased with increasing the hydrophobic EMA content [103]. The pH has a substantial influence in swelling properties and diffusion mechanism of acrylic-based materials. Hence, the swelling properties of semi-interpenetrating polymer networks of acrylamide-based polyurethanes decrease in acidic pH, whilst a reverse trend is observed in alkaline medium. However, these semi-IPNs are hydrolytically stable in phosphate buffer solution, which renders them a potential material in biomedicine and bioengineering [28]. Acrylic polymers which are pH-sensitive and biocompatible, such as poly(acrylic acid), are being used in many biomedical sectors [104]. Thus, they have attracted considerable interest mainly due to its therapeutic applications, because they possess great ability to swell reversibly with changes in pH. Functionalization of GO with PAA (GO-PAA) by in situ atom transfer radical polymerisation (ATRP) has shown great potential as intracellular protein carriers [105]. Poly(acryl amide-co-2-acrylamido-2-methyl-1-propanesulfonic acid-co-acrylamido glycolic acid) is a pH-sensitive terpolymer hydrogel suitable for drug release, which has shown a quasi-Fickian diffusion behaviour. These hydrogels demonstrate a sharp change in its water sorption and molecular weight between cross-links of the network with a change in pH of the swelling media [106]. The effect of temperature on swelling properties of acrylic hydrogels such as the thermosensitive poly(N-isopropyl acrylamide-co-acrylic acid) hydrogels is also very important [103], and they can be modified to exhibit fast temperature sensitivity and enhanced oscillating swelling-deswelling properties [107].

6. Antimicrobial capacity

Microbial infections are of serious concern and often lead to implant failure, which may cause major economic losses and suffering among patients despite the use of antibiotics and the aseptic processing conditions. Therefore, novel antimicrobial materials and strategies are urgently needed in biomedicine and bioengineering [108]. Nevertheless, acrylics itself do not possess antimicrobial activity intrinsically, and therefore some fillers and antimicrobial agents need to be incorporated usually by physical blending in order to produce an acrylic-based material able to treat and/or impede microbial infections [109]. Thus, graphene has emerged as a novel green broad-spectrum antimicrobial material with tolerable cytotoxicity in mammalian cells. Its antimicrobial action has been reported to be produced via physical damages through direct contact of its sharp edges with microbial membranes and destructive extraction of lipid molecules. Thus, graphene-based nanocomposites can be used in a broad range of biomedical applications due to its superior antimicrobial properties and good biocompatibility [110]. Methyl methacrylate was firstly used in tooth restoration in 1937, thereby indicating methacrylate monomers on polymerisation were biocompatible and have been extensively used as dental materials such as denture bases, adhesives, tissue conditioners, etc. [109]. The most popular methacrylate monomers employed in commercial dental resin-based materials are methyl methacrylate (MMA), 2,2-bis[4-(2-hydroxy-3-methacryloyloxypropyl)-phenyl]propane (Bis-GMA), 1,6-bis-[2-methacryloyloxyethoxycarbonylam ino]-2,4,4-trimethylhexane (UDMA) and tri-ethylene glycol dimethacrylate (TEGDMA) [111]. Nevertheless, these polymers are not inherently antimicrobial in nature; thus, the strategy of designing acrylic hydrogels with desired antimicrobial performance is important. Silver nanoparticles (Ag NPs) exhibit strong antibacterial activity against *Escherichia coli*, and thus composites with AgNPs have been synthesised to not only confer antimicrobial properties but also improve the mechanical performance of acrylic resins for dental applications. The release of silver ions upon immersion of the dental composite in water produces antimicrobial effect with negligible toxicity to humans [112]. Antimicrobial PMMA nanofibres with incorporated silver nanoparticles have been prepared by radical-mediated dispersion polymerisation and showed superior antimicrobial activity than silver sulfadiazine and silver nitrate at the same silver concentration [113]. Unfortunately, highly undesired infections are also frequent after orthopaedic procedures. In addition, the growing increase of antibiotic resistance is progressively decreasing the efficacy of such medical treatment. Thus, in this regard, the integration of powerful antimicrobial silver nanoparticles in polymeric acrylic-based nanocomposites is also an alternative antimicrobial approach [114]. The combination of previous strategies (graphene and Ag NPs) to develop antimicrobial hydrogels with good water maintaining ability is of particular significance for wound healing applications. Thus, a series of hydrogels have been prepared by cross-linking of Ag/graphene composites with acrylic acid and N,N′-methylene bisacrylamide at different compositions. In that study, hydrogels with an optimal Ag to graphene mass ratio of 5:1 were prepared and exhibited simultaneously excellent biocompatibility, much better antimicrobial properties than other hydrogels, high swelling ratio and good extensibility. Furthermore, in vivo experiments demonstrated that this nanocomposite hydrogel could significantly accelerate the healing rate of artificial wounds in rats, and it helped to successfully reconstruct an intact and thickened epidermis during 15 days of healing of impaired wounds [115]. In the same way, grafting of acrylic acid (AA) onto poly(ethylene terephthalate) (PET) films by gamma ray-induced graft copolymerization with silver nanoparticles showed strong and stable antimicrobial activity [116].

7. Acrylic-based scaffolds for tissue engineering

In regenerative medicine, regeneration and repair of diseased tissues can be addressed by tissue engineering approaches. A significant amount of research has focussed on the development of porous scaffolds that can interact directly with the seeded or laden cells. The porous 3D scaffolds provide the pathway for cell growth, proliferation and differentiation. The properties of the scaffolds are not only dependent on the base composition, but the structural and morphological features influence cell-material interaction. Hydrogels have emerged as leading candidates for engineered tissue scaffolds due to their biocompatibility and similarities to the native extracellular matrix. However, precise control of hydrogel properties, such as high porosity and tuning of mechanical properties, remains a challenge. Bulk porosities in polymers have been created by traditional techniques and have demonstrated success in hydrogels for tissue engineering applications. However, some problems related to direct cell encapsulation often occur. Nonetheless, a broad range of emerging technologies have demonstrated the ability to control porosity and its morphology in hydrogels, producing engineered tissues with structure and function similar to native tissues [117]. Ideal scaffolds should mimic the natural extracellular matrix (ECM), and a balance between temporary mechanical function with mass transport to aid biological delivery and tissue regeneration is of paramount importance. The interconnection and geometry of pores depend primarily on the tissue to be regenerated which then provide information on the desired physicochemical properties and mechanical resistance of the material. Several methods and techniques have been reported to produce *scaffolds*: gas foaming [118], fibre meshes sintering [119], solvent casting [120], polymerisation in solution [101, 121, 122], porogen leaching method [123, 124], freeze-drying methods [125, 126], electrospinning [127], 3D printing [128], 3D bioplotting of scaffold and cells [129], etc. For instance, acrylic scaffolds with interconnected spherical pores and controlled hydrophilicity were synthesised using a template of sintered PMMA microspheres of controlled size. In these scaffolds, their pore size, their connectivity, their porosity and their physicochemical properties could be tailored in an independent way by copolymerization of hydrophobic ethyl acrylate and hydrophilic hydroxyethyl methacrylate comonomers [124]. Four types of novel porous scaffolds of gelatin-PHEMA scaffolds were prepared with modulated architecture by freeze-drying technique and assessed by SEM and μ-CT. In this study, it is of note that the covalently bound gelatin sequences significantly enhanced the biocompatibility of the PHEMA-based hydrogels, which is very desirable for tissue engineering applications. Superporous acrylic scaffolds can be also prepared by the salt-leaching method using NaCl or $(NH4)_2SO_4$ as a porogen agent [130] or with many other porogenic compounds such as ammonium oxalate crystals [131]. Carbon dioxide (CO_2) subjected to supercritical conditions (P = 160–260 bar, T = 60°C) with rapid depressurization methods have also yielded porous architectures that are related to the supercritical parameters and polymer blend composition [132]. Scaffolds fabricated with CO_2 to create such porosity have received much attention in the past. Thus, highly porous (porosity greater than 85%) and well-interconnected scaffolds were reported in blends of poly(ethyl methacrylate) and tetrahydrofurfuryl methacrylate (PEMA/THFMA) showing promise for bioengineering applications in cartilage repair [133]. However, CO_2 processing of polymers can lead to porous scaffolds with limited interconnectivity between the pores. Other sophisticated techniques to produce highly porous supports are electrospinning, 3D printing and bioprinting. Thus, the electrospinning set-up consists of a high-voltage DC power supply, an infusion pumps and a syringe with a needle tip of usually 0.5 mm diameter. Thus, for example, biodegradable nanofibrous poly(L-lactic acid) (PLLA) scaffolds were fabricated in this method for tissue regeneration application [134]. 3D printing promises to produce reproducible complex biomedical devices with the help of computer design using patient-specific anatomical information. 3D printing has slowly evolved to produce one-of-a-kind devices, implants, scaffolds for tissue engineering and drug delivery systems among many other important uses. Nevertheless, 3D printing has still to overcome technical printing aspects such as the type of commercially printable materials available and the current slow speed of printing. The most common 3D printing technologies includes 3D printing, fused deposition modelling, selective laser sintering, stereolithography, and 3D plotting/direct-write/bioprinting and are still under research for their progress in tissue engineering. Bioprinting is currently considered the most advanced 3D printing technology because cells combined with custom 3D scaffolds are 3D printed for personalised regenerative medicine [128, 135, 136]. Mechanical resistance of scaffolds for tissue engineering applications depends on the material properties and on the interconnected pore morphology of the scaffold. This problem is more relevant in the case of porous acrylic hydrogels, which exhibit very weak mechanical performance in hydrated state [101]. Therefore, these porous structures often need to be reinforced. For example, hydrophilic scaffolds poly(2-hydroxyethyl acrylate) were reinforced in the form of nanocomposites with silica [9]. In addition, the silica phase of the scaffold was effectively interconnected and continuous, able to withstand pyrolysis without losing the pore architecture of the scaffold. Superporous scaffolds of PHEMA with cholesterol methacrylate (CHLMA) and laminin were developed in the presence of ammonium oxalate crystals to introduce interconnected superpores in the matrix that promoted cell-surface interaction [137]. PHEMA scaffolds have also been functionalized with laminin-derived Ac-CGGASIKVAVS-OH peptide sequences to promote cell adhesion and neural differentiation. With the same aim, nanofiber scaffolds produced by electrospinning were treated with oxygen plasma and then simultaneously in situ grafted with hydrophilic acrylic acid to obtain PLLA-g-PAA with a modified surface, which significantly enhanced cell adhesion and proliferation [134]. 3D microenvironment comprising fibronectin-coated PMMA/PC-based multicomponent polymeric systems has recently promoted the differentiation of primary human osteoblasts, which hereby renders a promising tool for tissue specific in vitro preconditioning of osteoblasts designated for clinically oriented bone augmentation or regeneration. In addition, morphogenesis and fluorescence dye-based live/dead staining revealed homogenous cell coverage of the microcavities. Nevertheless, azur II staining demonstrated formation of uniform sized multilayered aggregates, which exhibited progressive intracellular deposition of extracellular bone matrix constituents (fibronectin, osteocalcin and osteonectin) from day 7 on, and cells showed high viability up to 14 days [138]. Polymers based on an acrylate derivative of poly(ethylene glycol) (PEG) are currently being explored for applications in regenerative medicine. PEG is a hydrophilic, biocompatible and inert polymer that can be modified easily to alter the properties and manipulate both physical and biological properties. Acrylate derivatives such as poly(ethylene glycol) diacrylate and others are being explored to mimic native tissue environments [139]. Polysaccharide-based hydrogels have become increasingly studied as matrices in soft tissue engineering due to their known cytocompatibility [140]. Thus, for instance, cross-linkable dextran methacrylates and hyaluronan methacrylate hydrogel matrices have been proposed as excellent candidates for soft tissue reconstruction and have shown that their in vitro degradation behaviour can be controlled by the polysaccharide structure and the cross-linking density. In addition, under in vitro conditions, these biomaterials had no cytotoxic effects against fibroblasts, and the use of composite gels enhanced the adherence of cells [141]. Even though great advances are achieved in scaffold design of acrylic-based biomaterials, more research need to be carried out still in order to find new successful strategies capable of providing suitable advanced porous supports for tissue engineering applications.

8. Conclusions

Acrylic-based polymers possess versatile properties suitable for biomedical and bioengineering applications such as ophthalmology, orthopaedics, dentistry and tissue engineering. Nevertheless, the potential applications of this kind of polymers could be expanded exponentially if their chemical, physical and biological properties could be enhanced. In this regard, in the last few decades, acrylic-based materials have been produced in the form of interpenetrating polymer networks, composites with added particles such as fibres and nanocomposites with incorporated nanomaterials such as graphenes, in order to improve their mechanical, electrical, thermal, water sorption/diffusion, biological, antimicrobial and porosity properties, required for certain advanced applications. In addition, acrylic scaffolds have been fabricated with the suitable morphology for tissue regeneration by advanced fabrication methods. Nevertheless, in spite of the fabulous advancements presented in this chapter, many challenges still remain to tailor the properties of acrylate polymers for specific applications in the fields of biomedicine and bioengineering.

Acknowledgements

This work was supported by the 2019-231-003UCV grant from the Universidad Católica de Valencia *San Vicente Mártir*.

Author details

Ángel Serrano-Aroca[1]* and Sanjukta Deb[2]

1 Biomaterials and Bioengineering Lab, Centro de Investigación Traslacional San Alberto Magno, Universidad Católica de Valencia San Vicente Mártir, Valencia, Spain

2 Faculty of Dentistry, Oral and Craniofacial Sciences, King's College London, Guy's Hospital, London, UK

*Address all correspondence to: angel.serrano@ucv.es

References

[1] Chirila T, Harkin D. Biomaterials and Regenerative Medicine in Ophthalmology. 2nd ed. Amsterdam, The Netherlands: Elsevier; 2016

[2] Shalaby S, Nagatomi S, Peniston S. Polymeric biomaterials for articulating joint repair and total joint replacement. In: Polymers for Dental and Orthopedic Applications. Boca Raton, USA: CRC Press; 2007

[3] Van Blitterswijk C, De Boer J. Tissue Engineering. Oxford, United Kingdom: Academic Press; 2014

[4] Pratap B, Gupta RK, Bhardwaj B, Nag M. Resin based restorative dental materials: Characteristics and future perspectives. Japanese Dental Science Review. 2019;**55**:126-138

[5] Ratner BD, Hoffman AS, Schoen FJ, Lemons JE. Biomaterials Science: An Introduction to Materials in Medicine. Toronto, Canada: Academic Press; 2012

[6] Serrano-Aroca Á, Monleón-Pradas M, Gómez-Ribelles JL, Rault J. Thermal analysis of water in reinforced plasma-polymerised poly(2-hydroxyethyl acrylate) hydrogels. European Polymer Journal. 2015;**72**:523-534

[7] Serrano-Aroca Á, Monleón-Pradas M, Gómez-Ribelles JL. Plasma-induced polymerisation of hydrophilic coatings onto macroporous hydrophobic scaffolds. Polymer (Guildf). 2007;**48**(7):2071-2078

[8] Deb S. A review of improvements in acrylic bone cements. Journal of Biomaterials Applications. 1999;**14**(1): 16-47

[9] Rodríguez-Hernández JC, Serrano-Aroca Á, Gómez-Ribelles JL, Monleón-Pradas M. Three-dimensional nanocomposite scaffolds with ordered cylindrical orthogonal pores. Journal of Biomedical Materials Research Part B: Applied Biomaterials. 2008;**84**(2): 541-549

[10] Bisht A, Kumar V, Maity PC, Lahiri I, Lahiri D. Strong and transparent PMMA sheet reinforced with amine functionalized BN nanoflakes for UV-shielding application. Composites. Part B, Engineering. 2019;**176**:107274

[11] Serrano-Aroca Á. Latest improvements of acrylic-based polymer properties for biomedical applications. In: Acrylic Polymers in Healthcare. In, Rejika: InTechOpen; 2017. pp. 75-98

[12] De Lima GG, Elter JK, Chee BS, Magalhães WLE, Devine DM, Nugent MJD, et al. A tough and novel dual-response PAA/P(NiPAAM-co-PEGDMA) IPN hydrogels with ceramics by photopolymerization for consolidation of bone fragments following fracture. Biomedical Materials. 2019;**14**(5):054101

[13] Stoy V, Climent C. In Hydrogels: Speciality Plastics for Biomedical and Pharmaceutical Applications. Basel: Technomic Publishers; 1996

[14] Serrano-Aroca Á, Monleón-Pradas M, Gómez-Ribelles JL. Effect of crosslinking on porous poly(methyl methacrylate) produced by phase separation. Colloid & Polymer Science. 2008;**286**(2):209-216

[15] Ramaraj B, Radhakrishnan G. Modification of the dynamic swelling behaviour of poly(2-hydroxyethyl methacrylate) hydrogels in water through interpenetrating polymer networks (IPNs). Polymer (Guildf). 1994;**35**(10):2167-2173

[16] Danso R, Hoedebecke B, Whang K, Sarrami S, Johnston A, Flipse S, et al. Development of an oxirane/acrylate

interpenetrating polymer network (IPN) resin system. Dental Materials. 2018;**34**(10):1459-1465

[17] Gilbert JL, Ney DS, Lautenschlager EP. Self-reinforced composite poly (methy1 methacrylate): Static and fatigue properties. Biomaterials. 1995;**16**(14):1043-1055

[18] Serrano-Aroca Á, Gómez-Ribelles JL, Monleón-Pradas M, Vidaurre-Garayo A, Suay-Antón J. Characterisation of macroporous poly(methyl methacrylate) coated with plasma-polymerised poly(2-hydroxyethyl acrylate). European Polymer Journal. 2007;**43**(10):4552-4564

[19] Amerio E, Fabbri P, Malucelli G, Messori M, Sangermano M, Taurino R. Scratch resistance of nano-silica reinforced acrylic coatings. Progress in Organic Coatings. 2008;**62**(2):129-133

[20] Tahriri M, Del Monico M, Moghanian A, Tavakkoli Yaraki M, Torres R, Yadegari A, et al. Graphene and its derivatives: Opportunities and challenges in dentistry. Materials Science and Engineering C. 2019;**102**:171-185

[21] Kugler S, Kowalczyk K, Spychaj T. Progress in organic coatings hybrid carbon nanotubes/graphene modified acrylic coats. Progress in Organic Coatings. 2015;**85**:1-7

[22] Cha C, Shin SR, Gao X, Annabi N, Dokmeci MR, Tang XS, et al. Controlling mechanical properties of cell-laden hydrogels by covalent incorporation of graphene oxide. Small. 2014;**10**(3):514-523

[23] Faghihi S, Gheysour M, Karimi A, Salarian R. Fabrication and mechanical characterization of graphene oxide-reinforced poly (acrylic acid)/gelatin composite hydrogels. Journal of Applied Physics. 2014;**115**:083513

[24] Sánchez-Correa F, Vidaurre-Agut C, Serrano-Aroca A, Campillo-Fernández AJ. Poly(2-hydroxyethyl acrylate) hydrogels reinforced with graphene oxide: Remarkable improvement of water diffusion and mechanical properties. Journal of Applied Polymer Science. 2017;**135**(15):46158

[25] Ha H, Shanmuganathan K, Ellison CJ. Mechanically stable thermally crosslinked poly(acrylic acid)/reduced graphene oxide aerogels. ACS Applied Materials & Interfaces. 2015;**7**(11):6220-6229

[26] Sperling LH, Mishra V. The current status of interpenetrating polymer networks. Polymers for Advanced Technologies. 1996;7(4):197-208

[27] Maity J, Ray SK. Enhanced adsorption of methyl violet and Congo red by using semi and full IPN of polymethacrylic acid and chitosan. Carbohydrate Polymers. 2014;**104**(1):8-16

[28] Merlin DL, Sivasankar B. Synthesis and characterization of semi-interpenetrating polymer networks using biocompatible polyurethane and acrylamide monomer. European Polymer Journal. 2009;**45**(1):165-170

[29] Wang M, Pramoda KP, Goh SH. Mechanical behavior of pseudo-semi-interpenetrating polymer networks based on double-C 60 -end-capped poly (ethylene oxide) and poly (methyl methacrylate). Chemistry of Materials. 2004;**16**:3452-3456

[30] Hua F, Hu C. Interpenetrating polymer networks of epoxy resin and urethane acrylate resin 2. Morphology and mechanical property. European Polymer Journal. 2000;**36**(1):27-33

[31] Chakrabarty D, Das B, Roy S. Epoxy resin – Poly (ethyl methacrylate) interpenetrating polymer networks: Morphology. Mechanical. Journal of Applied Polymer Science. 1998;**67**(6):1051-1059

[32] Vabrik R, Czajlik I, Tury G, Rusznak I, Ille A, Vig A. A study of epoxy resin-acrylated polyurethane semi-interpenetrating polymer networks. Journal of Applied Polymer Science. 1998;**68**(1):111-119

[33] Dragan ES. Design and applications of interpenetrating polymer network hydrogels. A review. Chemical Engineering Journal. 2014;**243**:572-590

[34] Lee J, Kim S, Kim S, Lee Y, Lee K. Synthesis and characteristics of interpenetrating polymer network hydrogel composed of chitosan and poly (acrylic acid). Journal of Applied Polymer Science. 1999;**73**(1):113-120

[35] Groover MP. Fundamentals of Modern Manufacturing Materials, Processes, and Systems 4th ed. New York, USA: John Wiley & Sons; 2010

[36] Stipho HD. Repair of acrylic resin denture base reinforced with glass fiber. The Journal of Prosthetic Dentistry. 1998;**80**(5):546-550

[37] Young CD, Wu JR, Tsou TL. High-strength, ultra-thin and fiber-reinforced pHEMA artificial skin. Biomaterials. 1998;**19**(19):1745-1752

[38] Saheb DN, Jog JP. Natural fiber polymer composites: A review. Advances in Polymer Technology. 1999;**18**(4):351-363

[39] Ku H, Wang H, Pattarachaiyakoop N, Trada M. A review on the tensile properties of natural fiber reinforced polymer composites. Composites. Part B, Engineering. 2011;**42**(4):856-873

[40] Chen C, Li D, Hu Q, Wang R. Properties of polymethyl methacrylate-based nanocomposites: Reinforced with ultra-long chitin nanofiber extracted from crab shells. Materials and Design. 2014;**56**:1049-1056

[41] Dhas N, Parekh K, Pandey A, Kudarha R, Mutalik S, Mehta T. Two dimensional carbon based nanocomposites as multimodal therapeutic and diagnostic platform: A biomedical and toxicological perspective. Journal of Controlled Release. 2019;**308**:130-161

[42] Kokubo T. Design of bioactive bone substitutes based on biomineralization process. Materials Science and Engineering: C. 2005;**25**(2):97-104

[43] Hench LL, West JK. The sol-gel process. Chemical Reviews. 1990;**90**(1):33-72

[44] Trakulsujaritchok T, Hourston DJ. Damping characteristics and mechanical properties of silica filled PUR/PEMA simultaneous interpenetrating polymer networks. European Polymer Journal. 2006;**42**(11):2968-2976

[45] Frígols B, Martí M, Salesa B, Hernández-Oliver C, Aarstad O, Ulset AST, et al. Graphene oxide in zinc alginate films: Antibacterial activity, cytotoxicity, zinc release, water sorption/diffusion, wettability and opacity. Ng DYW, PLoS One. 2019;**14**(3):e0212819

[46] Salesa B, Martí M, Frígols B, Serrano-Aroca Á. Carbon nanofibers in pure form and in calcium alginate composites films: New cost-effective antibacterial biomaterials against the life-threatening multidrug-resistant Staphylococcus epidermidis. Polymers (Basel). 2019;**11**(3):453. Available from: https://www.mdpi.com/2073-4360/11/3/453

[47] Rivera-Briso AL, Aachmann FL, Moreno-Manzano V, Serrano-Aroca Á. Graphene oxide nanosheets versus carbon nanofibers: Enhancement of physical and biological properties of poly(3-hydroxybutyrate-co-3-hydroxyvalerate) films for biomedical applications. International Journal of Biological Macromolecules. 2019;**143**:1000-1008

[48] Serrano-Aroca Á, Deb S. Synthesis of irregular graphene oxide tubes using green chemistry and their potential use as reinforcement materials for biomedical applications. PLoS One. 2017;**12**(9):e0185235

[49] Serrano-Aroca Á, Iskandar L, Deb S. Green synthetic routes to alginate-graphene oxide composite hydrogels with enhanced physical properties for bioengineering applications. European Polymer Journal. 2018;**103**:198-206

[50] Martí M, Frígols B, Salesa B, Serrano-Aroca Á. Calcium alginate/graphene oxide films: Reinforced composites able to prevent Staphylococcus aureus and methicillin-resistant Staphylococcus epidermidis infections with no cytotoxicity for human keratinocyte HaCaT cells. European Polymer Journal. 2019;**110**:14-21

[51] Llorens-Gámez M, Serrano-Aroca Á. Low-cost advanced hydrogels of calcium alginate/carbon nanofibers with enhanced water diffusion and compression properties. Polymers (Basel). 2018;**10**(4):405

[52] Rivera-Briso AL, Serrano-Aroca Á. Poly(3-Hydroxybutyrate-co-3-Hydroxyvalerate): Enhancement strategies for advanced applications. Polymers. 2018;**10**:732

[53] Serrano-Aroca Á, Ruiz-Pividal J-F, Llorens-Gámez M. Enhancement of water diffusion and compression performance of crosslinked alginate films with a minuscule amount of graphene oxide. Scientific Reports. 2017;**7**(1):11684

[54] Novoselov KS, Geim AK, Morozov SV, Jiang D, Zhang Y, Dubonos SV, et al. Electric field effect in atomically thin carbon films. Science. 2004;**306**(5696):666-669

[55] Geim AK, Novoselov KS. The rise of graphene. Nature Materials. 2007;**6**:183-191

[56] Balandin AA, Ghosh S, Bao W, Calizo I, Teweldebrhan D, Miao F, et al. Superior thermal conductivity of single-layer graphene. Nano Letters. 2008;**8**(3):902-907

[57] Lee C, Wei X, Kysar JW, Hone J. Measurement of the elastic properties and intrinsic strength of monolayer graphene. Science. 2008;**321**(5887):385-388

[58] Kalbacova M, Broz A, Kong J, Kalbac M. Graphene substrates promote adherence of human osteoblasts and mesenchymal stromal cells. Carbon. 2010;**48**(15):4323-4329

[59] Zhan Y, Meng F, Yang X, Liu X. Magnetite-graphene nanosheets (GNs)/poly(arylene ether nitrile) (PEN): Fabrication and characterization of a multifunctional nanocomposite film. Colloids and Surfaces A: Physicochemical and Engineering Aspects. 2011;**390**(1-3):112-119

[60] Dong J, Ding J, Weng J, Dai L. Graphene enhances the shape memory of poly (acrylamide-co-acrylic acid) grafted on graphene. Macromolecular Rapid Communications. 2013;**34**(8):659-664

[61] Shen J, Yan B, Li T, Long Y, Li N, Ye M. Study on graphene-oxide-based polyacrylamide composite hydrogels. Composites. Part A, Applied Science and Manufacturing. 2012;**43**(9):1476-1481

[62] Huang Y, Zeng M, Ren J, Wang J, Fan L, Xu Q. Preparation and swelling properties of graphene oxide/poly(acrylic acid-co-acrylamide) super-absorbent hydrogel nanocomposites. Colloids and Surfaces A: Physicochemical and Engineering Aspects. 2012;**401**:97-106

[63] Shen J, Yan B, Li T, Long Y, Li N, Ye M. Mechanical, thermal and swelling properties of poly(acrylic

acid)-graphene oxide composite hydrogels. Soft Matter. 2012;**8**(6):1831-1836

[64] Potts JR, Lee SH, Alam TM, An J, Stoller MD, Piner RD, et al. Thermomechanical properties of chemically modified graphene/ poly(methyl methacrylate) composites made by in situ polymerization. Carbon. 2011;**49**(8):2615-2623

[65] Zeng J, Saltysiak B, Johnson WS, Schiraldi DA, Kumar S. Processing and properties of poly(methyl methacrylate)/carbon nano fiber composites. Composites. Part B, Engineering. 2004;**35**(2):173-178

[66] Kunanuruksapong R, Sirivat A. Electrical properties and electro-mechanical responses of acrylic elastomers and styrene copolymers: Effect of temperature. Applied Physics A: Materials Science & Processing. 2008;**92**(2):313-320

[67] Zhou C, Wu Q. A novel polyacrylamide nanocomposite hydrogel reinforced with natural chitosan nanofibers. Colloids and Surfaces. B, Biointerfaces. 2011;**84**(1):155-162

[68] Iwamoto S, Nakagaito AN, Yano H, Nogi M. Optically transparent composites reinforced with plant fiber-based nanofibers. Applied Physics A: Materials Science & Processing. 2005;**81**(6):1109-1112

[69] Meneghetti P, Qutubuddin S. Synthesis, thermal properties and applications of polymer-clay nanocomposites. Thermochimica Acta. 2006;**442**(1-2):74-77

[70] Medeiros AMMS, Machado F, Rubim JC. Synthesis and characterization of a magnetic bio-nanocomposite based on magnetic nanoparticles modified by acrylated fatty acids derived from castor oil. European Polymer Journal. 2015;**71**:152-163

[71] Rodrigues JR, Alves NM, Mano JF. Nacre-inspired nanocomposites produced using layer-by-layer assembly: Design strategies and biomedical applications. Materials Science and Engineering C. 2017;**76**:1263-1273

[72] Shahin M, Munir K, Wen C, Li Y. Magnesium matrix nanocomposites for orthopedic applications: A review from mechanical, corrosion, and biological perspectives. Acta Biomaterialia. 2019;**96**:1-19

[73] Kumar S, Sarita NM, Dilbaghi N, Tankeshwar K, Kim KH. Recent advances and remaining challenges for polymeric nanocomposites in healthcare applications. Progress in Polymer Science. 2018;**80**:1-38

[74] Wang L, Zhang Y, Wu A, Wei G. Designed graphene-peptide nanocomposites for biosensor applications: A review. Analytica Chimica Acta. 2017;**985**:24-40

[75] Zhao M, Forrester JV, McCaig CD. A small, physiological electric field orients cell division. Proceedings of the National Academy of Sciences of the United States of America. 1999;**96**:4942-4946

[76] Yao L, Shanley L, Mccaig C, Zhao M. Small applied electric fields guide migration of hippocampal neurons. Journal of Cellular Physiology. 2008;**216**(2):527-535

[77] Woo DG, Shim MS, Park JS, Yang HN, Lee DR, Park KH. The effect of electrical stimulation on the differentiation of hESCs adhered onto fibronectin-coated gold nanoparticles. Biomaterials. 2009;**30**(29):5631-5638

[78] Sisken BF, Walker J, Orgel M. Prospects on clinical applications of electrical stimulation for nerve regeneration. Journal of Cellular Biochemistry. 1993;**51**(4):404-409

[79] Zhao M. Electrical fields in wound healing-An overriding signal that directs cell migration. Seminars in Cell & Developmental Biology. 2009;**20**(6):674-682

[80] Kirson ED, Dbalý V, Tovaryš F, Vymazal J, Soustiel JF, Itzhaki A, et al. Alternating electric fields arrest cell proliferation in animal tumor models and human brain tumors. Proceedings of the National Academy of Sciences. 2007;**104**(24):10152-10157

[81] Fabbri P, Valentini L, Bittolo Bon S, Foix D, Pasquali L, Montecchi M, et al. In-situ graphene oxide reduction during UV-photopolymerization of graphene oxide/acrylic resins mixtures. Polymer (Guildf). 2012;**53**(26):6039-6044

[82] Huang YL, Tien HW, Ma CC, Yang SY, Wu SY, Liu HY, et al. Effect of extended polymer chains on properties of transparent graphene nanosheets conductive film. Journal of Materials Chemistry. 2011;**21**(45):18236

[83] Huang P, Chen W, Yan L. An inorganic-organic double network hydrogel of graphene and polymer. Nanoscale. 2013;**5**(13):6034-6039

[84] Iijima S. Helical microtubules of graphitic carbon. Nature. 1991;**354**(6348):56-58

[85] Rao CNR, Satishkumar BC, Govindaraj A. Nanotubes. ChemPhysChem. 2001;**2**:78-105

[86] Wang ZG, Wang Y, Xu H, Li G, Xu ZK. Carbon nanotube-filled nanofibrous membranes electrospun from poly(acrylonitrile-co-acrylic acid) for glucose biosensor. Journal of Physical Chemistry C. 2009;**113**(7):2955-2960

[87] Logakis E, Pandis CH, Pissis P, Pionteck J, Pötschke P. Highly conducting poly(methyl methacrylate)/carbon nanotubes composites: Investigation on their thermal, dynamic-mechanical, electrical and dielectric properties. Composites Science and Technology. 2011;**71**(6):854-862

[88] Huang ZH, Qiu KY. The effects of interactions on the properties of acrylic polymers/silica hybrid materials prepared by the in situ sol-gel process. Polymer (Guildf). 1997;**38**(3):521-526

[89] Ramanathan T, Abdala AA, Stankovich S, Dikin DA, Herrera-Alonso M, Piner RD, et al. Functionalized graphene sheets for polymer nanocomposites. Nature Nanotechnology. 2008;**3**(6):327-331

[90] Laachachi A, Cochez M, Ferriol M, Lopez-Cuesta JM, Leroy E. Influence of TiO2 and Fe2O3 fillers on the thermal properties of poly(methyl methacrylate) (PMMA). Materials Letters. 2005;**59**(1):36-39

[91] Laachachi A, Leroy E, Cochez M, Ferriol M, Lopez Cuesta JM. Use of oxide nanoparticles and organoclays to improve thermal stability and fire retardancy of poly(methyl methacrylate). Polymer Degradation and Stability. 2005;**89**(2):344-352

[92] Aymonier C, Bortzmeyer D, Thomann R, Muelhaupt R. Poly (methyl methacrylate)/palladium nanocomposites: Synthesis and characterization of the morphological, thermomechanical, and thermal properties. Chemistry of Materials. 2003;**15**(25):4874-4878

[93] Roorda WE, Bouwstra JA, de Vries MA, Junginger HE. Thermal analysis of water in p(HEMA) hydrogels. Biomaterials. 1988;**9**(6):494-499

[94] Tanaka M, Mochizuki A. Effect of water structure on blood compatibility--thermal analysis of water in poly(meth)acrylate. Journal of Biomedical Materials Research. Part A. 2004;**68**(4):684-695

[95] Serrano-Aroca Á, Campillo-Fernández AJ, Gómez-Ribelles JL, Monleón-Pradas M, Gallego-Ferrer G, Pissis P. Porous poly(2-hydroxyethyl acrylate) hydrogels prepared by radical polymerisation with methanol as diluent. Polymer (Guildf). 2004;**45**(26):8949-8955

[96] Clayton AB, Chirila TV, Lou X. Hydrophilic sponges based on 2-Hydroxyethyl methacrylate. V . Effect of crosslinking agent reactivity on mechanical properties. Polymer International. 1997;**44**:201-207

[97] Monleón-Pradas M, Gómez-Ribelles JL, Serrano-Aroca Á, Gallego-Ferrer G, Suay-Antón J, Pissis P. Interaction between water and polymer chains in poly(hydroxyethyl acrylate) hydrogels. Colloid & Polymer Science. 2001;**279**(4):323-330

[98] Ahmed EM. Hydrogel: Preparation, characterization, and applications - A review. Journal of Advanced Research. 2015;**6**(2):105-121

[99] Tanaka T. Gels. Scientific American. 1981;**244**(1):124-136 138

[100] Crank J. The Mathematics of Diffusion. 2nd Ed. Bristol, United Kingdom: Oxford University Press; 1975

[101] Monleón-Pradas M, Gómez-Ribelles JL, Serrano-Aroca Á, Gallego Ferrer G, Suay Antón J, Pissis P. Porous poly (2-hydroxyethyl acrylate) hydrogels. Polymer (Guildf). 2001;**42**(10):4667-4674

[102] Gehrke SH, Biren D, Hopkins JJ. Evidence for Fickian water transport in initially glassy poly(2-hydroxyethyl methacrylate). Journal of Biomaterials Science. Polymer Edition. 1995;**6**(4):375-390

[103] Wang J, Wu W. Swelling behaviors, tensile properties and thermodynamic studies of water sorption of 2-hydroxyethyl methacrylate/epoxy methacrylate copolymeric hydrogels. European Polymer Journal. 2005;**41**(5):1143-1151

[104] Zhao Y, Kang J, Tan T. Salt-, pH- and temperature-responsive semi-interpenetrating polymer network hydrogel based on poly(aspartic acid) and poly(acrylic acid). Polymer (Guildf). 2006;**47**(22):7702-7710

[105] Kavitha T, Kang IK, Park SY. Poly(acrylic acid)-grafted graphene oxide as an intracellular protein carrier. Langmuir. 2014;**30**(1):402-409

[106] Krishna-Rao KSV, Ha CS. PH sensitive hydrogels based on acryl amides and their swelling and diffusion characteristics with drug delivery behavior. Polymer Bulletin. 2009;**62**(2):167-181

[107] Zhang XZ, Yang YY, Wang FJ, Chung TS. Thermosensitive poly(N-isopropylacrylamide-co-acrylic acid) hydrogels with expanded network structures and improved oscillating swelling-deswelling properties. Langmuir. 2002;**18**(6):2013-2018

[108] Shi L, Chen J, Teng L, Wang L, Zhu G, Liu S, et al. The antibacterial applications of graphene and its derivatives. Small. 2016;**12**(31):4165-4184

[109] He J, Söderling E, Lassila LVJ, Vallittu PK. Incorporation of an antibacterial and radiopaque monomer in to dental resin system. Dental Materials. 2012;**28**(8):e110-e117

[110] Ji H, Sun H, Qu X. Antibacterial applications of graphene-based nanomaterials: Recent achievements and challenges. Advanced Drug Delivery Reviews. 2016;**105**:176-189

[111] Moszner N, Salz U. New developments of polymeric dental

composites. Progress in Polymer Science. 2001;**26**(4):535-576

[112] Guo C, Zhou L, Jianxiong L. Effects of expandable graphite and modified ammonium polyphosphate on the flame-retardant and mechanical properties of wood flour-polypropylene composites. Polymers and Polymer Composites. 2013;**21**(7):449-456

[113] Kong H, Jang J. Antibacterial properties of novel poly(methyl methacrylate) nanofiber containing silver nanoparticles. Langmuir. 2008;**24**(5):2051-2056

[114] González-Sánchez MI, Perni S, Tommasi G, Morris NG, Hawkins K, López-Cabarcos E, et al. Silver nanoparticle based antibacterial methacrylate hydrogels potential for bone graft applications. Materials Science and Engineering: C. 2015;**50**:332-340

[115] Fan Z, Liu B, Wang J, Zhang S, Lin Q, Gong P, et al. A novel wound dressing based on Ag/graphene polymer hydrogel: Effectively kill bacteria and accelerate wound healing. Advanced Functional Materials. 2014;**24**(25):3933-3943

[116] Ping X, Wang M, Xuewu G. Surface modification of poly(ethylene terephthalate) (PET) film by gamma-ray induced grafting of poly(acrylic acid) and its application in antibacterial hybrid film. Radiation Physics and Chemistry. 2011;**80**(4):567-572

[117] Annabi N, Nichol JW, Zhong X, Ji C, Koshy S, Khademhosseini A, et al. Controlling the porosity and microarchitecture of hydrogels for tissue engineering. Tissue Engineering. Part B, Reviews. 2010;**16**(4):371-383

[118] Arora KA, Lesser AJ, McCarthy TJ. Compressive behavior of microcellular polystyrene foams processed in supercritical carbon dioxide. Polymer Engineering and Science. 1998;**38**(12):2055-2062

[119] Thomson RC, Wake MC, Yaszemski MJ, Mikos AG. Biodegradable polymer scaffolds to regenerate organs. Advances in Polymer Science. 1995;**122**:247-277

[120] Andrianova GP, Pakhomov SI. Porous materials from crystallizable polyolefins produced by gel technology. Polymer Engineering and Science. 1997;**37**(8):1367-1380

[121] Serrano-Aroca Á, Monleón-Pradas M, Gómez-Ribelles JL. Macroporous poly(methyl methacrylate) produced by phase separation during polymerisation in solution. Colloid & Polymer Science. 2007;**285**(7):753-760

[122] Serrano-Aroca Á, Llorens-Gámez M. Dynamic mechanical analysis and water vapour sorption of highly porous poly(methyl methacrylate). Polymer (Guildf). 2017;**125**:58-65

[123] Flynn L, Dalton PD, Shoichet MS. Fiber templating of poly (2-hydroxyethyl methacrylate) for neural tissue engineering. Biomaterials. 2003;**24**(23):4265-4272

[124] Brígido-Diego R, Pérez-Olmedilla M, Serrano-Aroca Á, Gómez-Ribelles JL, Monleón-Pradas M, Gallego-Ferrer G, et al. Acrylic scaffolds with interconnected spherical pores and controlled hydrophilicity for tissue engineering. Journal of Materials Science. Materials in Medicine. 2005;**40**(18):4881-4887

[125] Kang HW, Tabata Y, Ikada Y. Fabrication of porous gelatin scaffolds for tissue engineering. Biomaterials. 1999;**20**(14):1339-1344

[126] Dragusin DM, Van Vlierberghe S, Dubruel P, Dierick M, Van Hoorebeke L, Declercq HA, et al. Novel gelatin–PHEMA porous scaffolds for tissue engineering applications. Soft Matter. 2012;**8**(37):9589

[127] Agarwal S, Wendorff JH, Greiner A. Use of electrospinning

technique for biomedical applications. Polymer (Guildf). 2008;**49**(26):5603-5621

[128] Chia HN, Wu BM. Recent advances in 3D printing of tissue engineering scaffolds. Journal of Biological Engineering. 2015;**9**(4):2-14

[129] Derby B. Printing and prototyping of tissues and scaffolds. Science. 2012;**338**(6109):921-926

[130] Horák D, Hlídková H, Hradil J, Lapčíková M, Šlouf M. Superporous poly(2-hydroxyethyl methacrylate) based scaffolds: Preparation and characterization. Polymer (Guildf). 2008;**49**(8):2046-2054

[131] Kubinová Š, Horák D, Kozubenko N, Vaněček V, Proks V, Price J, et al. The use of superporous Ac-CGGASIKVAVS-OH-modified PHEMA scaffolds to promote cell adhesion and the differentiation of human fetal neural precursors. Biomaterials. 2010;**31**(23):5966-5975

[132] Velasco D, Benito L, Fernández-Gutiérrez M, San Román J, Elvira C. Preparation in supercritical CO2 of porous poly(methyl methacrylate)-poly(l-lactic acid) (PMMA-PLA) scaffolds incorporating ibuprofen. Journal of Supercritical Fluids. 2010;**54**(3):335-341

[133] Barry JJA, Silva MMCG, Cartmell SH, Guldberg RE, Scotchford CA, Howdle SM. Porous methacrylate tissue engineering scaffolds: Using carbon dioxide to control porosity and interconnectivity. Journal of Materials Science. 2006;**41**(13):4197-4204

[134] Park K, Hyun JJ, Kim JJ, Ahn KD, Dong KH, Young MJ. Acrylic acid-grafted hydrophilic electrospun nanofibrous poly(L-lactic acid) scaffold. Macromolecular Research. 2006;**14**(5):552-558

[135] Singh M, Jonnalagadda S. Advances in bioprinting using additive manufacturing. European Journal of Pharmaceutical Sciences. 2020;**143**:105167

[136] Bose S, Ke D, Sahasrabudhe H, Bandyopadhyay A. Additive manufacturing of biomaterials. Progress in Materials Science. 2018;**93**:45-111

[137] Kubinová Š, Horák D, Syková E. Cholesterol-modified superporous poly(2-hydroxyethyl methacrylate) scaffolds for tissue engineering. Biomaterials. 2009;**30**(27):4601-4609

[138] Altmann B, Steinberg T, Giselbrecht S, Gottwald E, Tomakidi P, Bächle-Haas M, et al. Promotion of osteoblast differentiation in 3D biomaterial micro-chip arrays comprising fibronectin-coated poly(methyl methacrylate) polycarbonate. Biomaterials. 2011;**32**(34):8947-8956

[139] Moore EM, West JL. Bioactive poly(ethylene glycol) acrylate hydrogels for regenerative engineering. Regenerative Engineering and Translational Medicine. 2019;**5**(2):167-179

[140] Salesa B, Llorens-Gámez M, Serrano-Aroca Á. Study of 1D and 2D carbon nanomaterial in alginate films. Nanomaterials. 2020;**10**(2):206

[141] Möller S, Weisser J, Bischoff S, Schnabelrauch M. Dextran and hyaluronan methacrylate based hydrogels as matrices for soft tissue reconstruction. Biomolecular Engineering. 2007;**24**(5):496-504

Chapter 5

Acrylic-Based Hydrogels as Advanced Biomaterials

Ángel Serrano-Aroca and Sanjukta Deb

Abstract

Acrylate based hydrogels are one of the most promising soft biocompatible material platforms that significantly contribute to the delivery of therapeutics, contact lenses, corneal prosthesis, bone cements and wound dressing, and are being explored widely for potential applications in the field of regenerative medicine. A significant number of these materials, which possess excellent water sorption properties, have been supported by the Food and Drug Administration (FDA) of the United States for different applications. Nonetheless, many of their physical and biological properties required for certain biomedical and bioengineering applications are often poor when they are in the hydrated state at the body temperature: tensile/compression performance, water diffusion, antimicrobial activity, antifouling capacity, biological response, porosity for the fabrication of supports or *scaffolds* for tissue engineering, electrical and/or thermal properties, among other properties. Consequently, new acrylic-based hydrogels have been designed as multicomponent systems such as interpenetrated polymer networks, composites and nanocomposite materials, which have exhibited superior properties able to substantially enhance potential uses of these materials in the biomedical and bioengineering industry.

Keywords: acrylic-based hydrogels, polymers, composites, nanocomposites, biomedicine, bioengineering

1. Introduction

Hydrogels are hydrophilic polymer networks that are capable of absorbing large amounts of water and retaining them, which make them a versatile class of materials especially for biomedical applications. The physical and biological properties largely depend on composition, polymerisation methods and crosslinking but in general are mechanically weak materials. Acrylic-based hydrogels are utilised in numerous fields of biomedicine and examples include therapeutic delivery [1], intraocular lenses, contact lenses and corneal prosthesis in ophthalmology [2], bone cements for orthopaedics [3], wound dressing [4], and tissue *scaffolds* for regenerative medicine [5], due to their diverse properties. Owing to their propensity to absorb large amounts of water or biological fluids, they tend to exhibit properties like the extracellular matrix. The ability to tune physical and mechanical properties of hydrogels and the inherent properties that facilitate diffusion of oxygen, biomolecules and waste metabolites make them a versatile group of polymers [6]. Many of these hydrogels have been supported by the US Food and Drug Administration (FDA) for differing applications. Nonetheless, their potential use in biomedical applications are sometimes limited by their low mechanical strength, biological interactions, electrical

IntechOpen

and/or thermal properties, water sorption and diffusion, antimicrobial and/or antifouling activity, porosity, etc. [7]. These shortcomings have led to the development of suitable advanced acrylic-based hydrogels and research is ongoing to find solutions. Thus, the approach of developing multicomponent polymeric systems or combination of materials and/or nanomaterials to form composites or nanocomposites with enhanced physical and biological properties is of interest.

2. Mechanical properties of hydrogels

Hydrogels form a versatile platform for a large number of biomedical applications; however, they are in general mechanically weak and improvement of these properties is one of the most desirable aims in the field of hydrogel engineering. Current research is focussing in this complex scientific field [8] especially as the mechanical integrity drops rapidly in the hydrated state. Thus, hydrogels have been reinforced through many established kinds of methods and techniques: forming block copolymers, in which hydrophobic and hydrophilic domains alternate [9], increasing crosslinking density [10], using binary systems composed of two or more mixed polymers as interpenetrating polymer networks [11], plasma grafting of a hydrogel onto a hydrophobic substrate [12–14], self-reinforced composite materials composed of fibres embedded in a matrix [15] and by reinforcement through sol-gel reactions [16]. Nevertheless, new procedures to enhance the mechanical performance of acrylics have been performed by incorporating 2D materials such as graphene (2010 Nobel Prize in Physics) [17] and other outstanding carbon-based materials such as carbon nanotubes (CNT) [18]. Graphene derivatives, such as graphene oxide (GO) [19–21] or reduced GO (rGO) [22], have also shown excellent reinforcement for acrylics, especially in the hydrated state, and for the enhancement of many other physical and biological properties.

2.1 Interpenetrating polymer networks

Interpenetrating polymer networks (IPNs) lead to reinforced polymer networks that enhance the mechanical properties of hydrogels. In 2003, the first double-network (DN) hydrogel with enhanced mechanical properties was reported [23]. This type of DN interwoven hydrogel architecture consisted of an interpenetrating polymer network (IPN) of a soft neutral polymer network within a more highly cross-linked network prepared by a two-step sequential free-radical polymerisation. This two-step chemical procedure consisted of synthesising a highly cross-linked polymer network, and subsequent swelling of this network in a water soluble monomer that was then polymerised inside. The second polymerisation step was conducted with or without the incorporation of a cross-linking agent. Therefore, an IPN is an advanced polymeric system that is often utilised in polymer engineering to enhance physical and biological properties by combination of functional properties. These multicomponent polymeric networks are composed of cross-linked polymers without covalent bonds between polymer networks. However, at least one of these networks is synthesised and/or cross-linked within the presence of the second network. In this field, six basic multicomponent polymeric morphologies can be distinguished: mechanic blends, graft copolymers, block copolymers, AB-cross-linked copolymers, semi-IPNs and full-IPNs [24]. Both polymer networks are cross-linked in a full-IPN [25], while an IPN having a polymer network embedded within the first cross-linked network produces a semi- or pseudo-IPN [26, 27]. IPNs can be produced while the two networks are synthesised at the same time as simultaneous interpenetrating polymer networks (SINs), or by swelling of the first polymer network into a solution containing the mixture of monomer, initiator and activator, usually with a cross-linker, as sequential IPNs. Thus, reinforced SINs and semi-SINs of hydrophilic poly(2-hydroxyethyl methacrylate) (PHEMA) networks have been synthesised with hydrophobic poly(ethylene glycol) polymer chains [28]. The sequential mode of synthesis has been employed to produce reinforced hydrogels as full IPNs and semi-IPNs of weak gelatin with polyacrylic acid (PAA) for studies of biological response in rats [29]. The mechanical properties of triple-network (TN) hydrogels synthesised from pseudo-SIPNs and pseudo-IPNs have exhibited that the presence of a loosely cross-linked third network modifies the mechanical performance of pseudo-SIPNs and pseudo-IPNs [30]. Many types of IPN hydrogels have been also developed with the goal of improving the mechanical behaviour and swelling/deswelling response [31]. For instance, IPN hydrogels of chitosan/poly(acrylic acid) (PAA) synthesised by UV radiation showed significant enhancement of mechanical properties, even in the hydrated state [32]. 'Smart' hydrogels possess the special property of being able to modify their volume/shape in response to small alterations of certain parameters of the surrounding ambient. These responsive hydrogels find application in numerous biomedical and bioengineering fields such as biological and therapeutic demands [33, 34] and sensing applications [35]. Encapsulation of cells for cartilage tissue engineering was reported using two biocompatible materials-agarose and poly(ethylene glycol) (PEG) diacrylate to form an IPN with superior mechanical integrity [36]. Under unconfined compression, these hydrogel networks were found to be fourfold higher in their shear modulus relative to a pure PEG-diacrylate network (39.9 vs. 9.9 kPa) and a 4.9-fold increase relative to a pure agarose network (8.2 kPa). In the field of hydrogel materials, advanced stimuli-responsive materials based on interpenetrating liquid crystal-hydrogel polymer networks have been engineered by combination of cholesteric liquid crystalline network that reflects colour and an interwoven poly(acrylic acid) network that provides humidity and pH response [37].

2.2 Acrylic-based composite hydrogels

Fibre reinforcement is known to enhance mechanical properties and the incorporation of fabrics produces significant mechanical improvement to polymer hydrogels. Thus, hydrogels consisting of a polymer matrix embedded with high-strength fibres, such as glass, aramid and carbon, have exhibited significant reinforcement [38]. In these kinds of materials, the mechanical properties are significantly enhanced and the biocompatibility of the acrylic polymer phase should remain unmodified. Thus, PHEMA hydrogels, which are one of the most studied hydrogel biomaterials, have been synthesised by incorporating various types of weaved and knitted fabrics and fibres, in order to enhance overall qualities in advanced biomedical wound dressing usage [39]. Nevertheless, in the last decades, natural fibres have attracted much interest as reinforcement agents for polymer-based composites due to their advantages over other type of conventional fibres such as those made from glass or carbon [40]. Thus, natural fibres such as flax, hemp, sisal, kenaf, jute, coir, kapok and banana, among many others, have shown many advantages over man-made glass and carbon fibres: lower cost, lower density, comparable specific tensile behaviour, less energy cost, non-abrasive, not irritating, lower health risk and sustainable properties such as renewability, biodegradability and recyclability [41]. Thus, natural ultra-long chitin natural fibres obtained from marine-river crab shell wastes have been added into PMMA resins to produce nanocomposites with outstanding properties for biomedical and bioengineering applications [42].

2.3 Acrylic-based nanocomposite hydrogels

Nanoparticles have generated significant interest and are a promising strategy of reinforcing hydrogels. Nanocomposites made with nanomaterials such as silica, graphene and its derivatives, nanofibres and various other nanoparticles have been reported for biomedical applications. Silica is a biocompatible material which has been reported to possess bioactive properties [43]. Silica possesses high biocompatibility and bioactive properties [43] and can also enhance the mechanical performance of hydrogels through filling or by the sol-gel process [44]. Thus, for example, biphasic matrices of hybrid acrylic-based nanocomposite hydrogels consisting of an organic phase of poly(2-hydroxyethyl acrylate) and an inorganic phase of silica network obtained by the sol-gel process of tetraethoxysilane (TEOS) showed improved physical properties [45]. Another reinforcement approach consists of combining IPNs with nanosilica filling. Thus, for instance, poly(acrylic acid) and alginate in the form of IPNs with the incorporation of nanosilica have shown improved compressive strength of the pure components [46]. On the other hand, the development of graphene-based nanocomposite hydrogels has exponentially increased during the last decade. Thus, graphene (GN) is a two-dimensional monolayer of sp2-bonded carbon atoms [47], which has attracted increasing interest owing to its excellent electrical and thermal conductivity [48, 49] and superior mechanical performance [50]. In addition, in the field of biomedicine, graphene is able to promote adherence of human osteoblasts and mesenchymal stromal cells [51]. The addition of small amounts of the oxidised form of GN, graphene oxide (GO), can significantly improve the mechanical strength of poly(2-hydroxyethyl acrylate) (PHEA) hydrogels [21]. Polyacrylamide (PAM) is generally a weak and brittle material; however, when reinforced with GO, it was reported to exhibit improved of mechanical properties [52]. GO is also a 2D nanomaterial with excellent physical properties [53] like graphene, usually produced from natural graphite that can be easily exfoliated into monolayer sheets. Nevertheless, GO is more utilised than GN in the synthesis of composite materials because it possesses hydrophilic oxygenated functional groups, such as hydroxyl (-OH), epoxy (-C-O-C-), carbonyl (-C=O) and carboxyl (-COOH), which render easier its dispersion in water [54–57]. Thus, the incorporation of GO nanosheets into poly(acrylic acid)/gelatin composite hydrogels significantly increased their Young's modulus and maximum stress. In addition, the hydrogel with GO (0.2% w/w)/PAA (20% w/w) showed the highest Young's modulus, whereas GO (0.2% w/w)/PAA (40% w/w) composites exhibited the highest maximum stress. These results suggest that GO nanosheets can be successfully used as reinforcing agents to improve the mechanical properties of hydrogel materials, which are often required for certain applications in tissue regeneration [20]. Multifunctional hydrogels with high mechanical performance, environmental stability, and dye-loading capacity has also been proposed as innovative synthetic approach for the 3D self-assembly of GO nanosheets and DNA [58]. Furthermore, the available oxygen-containing functional groups of GO have led to the synthesis of 3D cross-linked GO networks by coordination chemistry, as reinforcement micro-meter size carbon nanomaterials (CNMs) are able to enhance the mechanical performance of hydrogels, such as alginate, even more than single GO nanosheets [59]. Other CNMs such as carbon nanotubes (CNTs), discovered by Iijima [60], in the form of single wall carbon nanotubes (SWCNTs) and multi-wall carbon nanotubes (MWCNTs), as well as carbon nanofibres (CNFs) are being explored for the enhancement of mechanical and other physical and biological properties of hydrogels [61–68]. Plant fibre-based nanofibres are used as reinforcement agents to produce transparent hydrogels [69]. Novel PAM-based nanocomposite hydrogels produced with natural chitosan nanofibres via in situ free-radical polymerisation exhibited that this type of nanofibres can act simultaneously as a multifunctional cross-linker and as a reinforcing agent achieving higher compression strengths and storage modulus than those of the pure hydrogel [70]. Popular hydrogels such as polyvinyl alcohol (PVA) has been reinforced with nanoparticles of clay (usually 10 wt.% or less) for wound healing applications [71]. These polymer-clay nanocomposite hydrogels with uniformly dispersed inorganic particles with at least one dimension in the nanometre scale exhibited superior mechanical and thermal properties when compared to pure polymers or conventional composites [72].

3. Electrical properties

Stimulus responsive biomaterials are highly desirable in biomedicine and bioengineering. Electrical stimulation can regulate physiological activities such as cell division [73], migration [74], differentiation and death [75]. It has been reported that electrical stimulation helps both in spinal cord repair and cancer therapy. The ability to use electrically conducting polymers endogenously enables spatial control of stimulation [76–78]; hence, there is much focus on developing new conductive hydrogels for biomedical applications. Graphene is well-known for its excellent electrical conductivity [48]. However, its current cost is still very high. Therefore, more new composite materials are expected to be developed using reduced graphene oxide (rGO), which is produced from GO. GO cannot be utilised for the design of conductive composite hydrogels because it possesses very low electrical conductivity due to the oxygen-containing functional groups located at the basal planes and edges. A recent paper reported a single-step procedure starting from a homogeneous water dispersion of GO, to form rGO during photopolymerisation of a resin induced by UV radiation [79]. Grafting of poly(acryl amide)/poly(acrylic acid) onto the surface of GO followed by a reduction to rGO nanosheets by a two-step chemical reduction with increased conductivity has been performed in order to fabricate transparent conductive films [80]. Advanced conductive DN hydrogels composed of rGO and poly(acrylic acid) have been synthesised by a two-step procedure with a reduction-induced in situ self-assembly [81]. A nacre-inspired nanocomposite of rGO and PAA prepared via a vacuum-assisted filtration self-assembly process exhibited abundant hydrogen bonding between GO and PAA that resulted in both high strength and toughness, which are higher than that of pure reduced GO. Moreover, this composite also displayed high electrical conductivity with potential in many biomedical applications such as flexible electrodes and artificial muscles. Carbon nanotubes (CNTs), on the other hand, are being exploited for biosensing units because of their excellent electrical properties with a superb conductivity and remarkable mechanical properties [82]. Ultrasensitive electrochemical biosensors have been developed with CNTs because of their unique electrical properties. Glucose biosensors for diabetics have been developed with nanofibrous membranes filled with multiwalled carbon nanotubes (MWCNTs) electrospun from mixtures of poly(acrylonitrile-co-acrylic acid) (PANCAA) and MWCNT [83]. Dielectrophoretically aligned carbon nanotubes were proposed to control the electrical and mechanical properties of gelatin methacrylate (GelMA) hydrogels [84]. In addition, in the field of biomedicine, these GelMA-based hydrogels exhibited excellent maturation of contractile muscle cells.

4. Thermal properties

Hydrogels placed in the human body do not need to endure temperatures higher than that of body temperature; however, it is of importance to have an understanding of the thermal properties and improvements can enhance its long-term performance. For instance, semi-IPNs of polyurethane incorporated into a polyacrylamide network showed improved thermal properties [26]. Differential scanning calorimetry of PHEMA/SiO_2 hybrids exhibited two glass transition temperatures (*Tg*). In addition, the SiO_2 content have shown to be able to modify the *Tg* shift of the thermal transition [85]. However, polymer nanocomposites with functionalized graphene sheets (FGNSs) showed a *Tg* shift of up to 40 and 30°C in poly(acrylonitrile) with the addition of 1% *w/w* of this carbon-based material [86]. Another successful enhancement of thermal properties of hydrogels, in terms of thermal behaviour and degradation, can be achieved by incorporating nanoparticle fillers. Thus, thermally stable, soft and magnetic field-driven actuators with muscle-like flexible PHEMA-based hydrogels were prepared with crosslinking metal nanoparticles added into the polymer backbone [87]. In this research line, the mechanical and thermal performance of renewable and biocompatible hydrogels of gelatin has been enhanced through cross-linking with cellulose nanowhiskers [88]. Acrylic-based hydrogels are hydrophilic polymers that are able to absorb large amounts of water in biomedicine and bioengineering due to their contact with fluids in cells or tissue in the human body. Therefore, the thermal analysis of water and its influence on the hydrated hydrogel properties becomes essential in this field. In this regard, many studies have been reported with acrylic hydrogels such as PHEMA [89], bulk and plasma-poly(2-hydroxyethyl acrylate) (*pl*PHEA) [14].

5. Water sorption/diffusion

The behaviour of water in hydrogels is also very important for any biomedical applications because properties such as water sorption and water diffusion play a very important role in cell survival, especially relevant in tissue regeneration [5]. Thus, acrylic hydrogels such as PHEMA or PHEA are important due to their unique properties of hydrophilicity, swelling and deswelling [90–92]. Their excellent water sorption behaviour has rendered this type of hydrophilic materials very promising for a wide range of biomedical and bioengineering applications such as wound healing, controlled drug delivery, regenerative medicine, etc. [6, 93]. Their hydrophilic functional groups attached to the polymeric backbone provide the ability to absorb water, while their resistance to dissolution arises from cross-linking of polymer chains [94]. Nevertheless, these single-network hydrogels possess very weak mechanical properties and slow swelling response. Therefore, reinforcement of hydrogels is deemed necessary to exponentially increase their potential applications in biomedicine and bioengineering. However, the enhancement of mechanical properties can significantly affect the water sorption and diffusion of polymer hydrogels. Thus, reinforcement approaches combining hydrophilic and hydrophobic functional groups as multicomponent polymeric systems can yield to a decrease of water sorption. The enhancement of mechanical properties of hydrogels through the addition of GO nanosheets can also modify their water sorption performance. Thus, for instance, the swelling rates of graphene oxide/poly(acrylic acid-co-acrylamide) nanocomposite hydrogels increased with increasing GO content to about 0.30% *w/w* and then decreased with further increase in GO contents. It is worth noting that the hydrogel with only 0.10% *w/w* GO exhibited significant enhancement in swelling capacity in a neutral medium and could retain relatively higher swelling rates in acidic and basic solutions. Furthermore, a very low filler volume of GO can produce a very significant increase in water diffusion (almost 6 times faster) in cross-linked alginate [95]. Therefore, these GO-based super-absorbent hydrogels have potential applications in many fields such as biomedical engineering and hygiene products [54]. The water diffusion mechanisms [96] can also be altered by the modification of mechanical properties of hydrogels. Thus, poly(acrylic acid)-GO nanocomposite hydrogels, which are potential carriers for drug release, can be manipulated by changing the concentration of GO and tend to exhibit non-Fickian anomalous diffusion with decrease in deswelling ratio with increasing GO content [53]. Superabsorbent polymer hydrogels of sodium lignosulfonate-grafted poly(acrylic acid-co-acryl amide), synthesised by a ultrasonic method, also exhibited a non-Fickian water diffusion transport with a maximum water absorbency of 1350 $g{\cdot}g^{-1}$ [97]. There are many acrylic hydrogels, which have shown non-Fickian diffusion mechanism such as PHEA [92, 98]. Nevertheless, a well-known water-swellable biomedical polymer hydrogel such as PHEMA has shown to be governed by Fickian diffusion, even though water sorption is not classically Fickian [99]. Thus, advanced hydrogels based on 2-hydroxyethyl methacrylate (HEMA) and epoxy methacrylate (EMA) produced via bulk polymerisation exhibited also a Fickian swelling behaviour, and the equilibrium water content (EWC) decreased with increasing the hydrophobic EMA content as expected [100]. The pH of the environment influences also the swelling capacity and diffusion mechanism of acrylic-based hydrogels. Hence, for example, the swelling properties of semi-IPNs of acrylamide-based polyurethanes decreased in acidic pH, while a reverse trend was observed in alkaline medium. However, these semi-IPNs were hydrolytically stable in phosphate buffer solution, which render them potential hydrogel materials for biomedical and bioengineering applications [26]. PAA is a pH-sensitive and biocompatible polymer that is being used in a wide range of biomedical fields [34] and has attracted considerable interest due to its capacity to swell reversibly with changes in pH. Thus, the functionalization of GO nanosheets with PAA (GO-PAA) by in situ atom transfer radical polymerisation (ATRP) have demonstrated great potential as intracellular protein carriers using bovine serum albumin (BSA) [101]. This achievement is very important because proteins participate in all vital body processes and perform an essential function inside cells as enzymes, transduction signals and gene regulation. Poly(acryl amide-co-2-acrylamido-2-methyl-1-propanesulfonic acid-co-acrylamido glycolic acid) is another pH-sensitive terpolymer hydrogel suitable for drug release, which has shown a quasi-Fickian diffusion mechanism. These hydrogels showed a strong change of water sorption and molecular weight between cross-links of the network with a change in pH of the swelling media [102]. Temperature affects also very significantly the swelling properties of acrylic hydrogels [100]. Thus, for example, thermosensitive poly(N-isopropyl acrylamide-co-acrylic acid) hydrogels can be designed in order to exhibit fast temperature sensitivity and enhanced oscillating swelling-deswelling properties [103].

6. Antimicrobial and antifouling capacity

Microbial infections are becoming more and more serious because they often lead to implant failure, which may cause major economic losses and suffering among patients in spite of using antibiotics and aseptic conditions. Therefore, novel antimicrobial approaches are becoming more and more necessary in biomedicine in this antibiotics resistant era [104]. Thus, much effort is being made on the development of new advanced biomaterial hydrogels with high antimicrobial activity and non-toxic for human beings. Thus, antibacterial hydrogels of polydextran aldehyde and branched polyethylenimine have been prepared in the form of syringe-injectable bioadhesive [105]. These hydrogels have shown effective antibacterial activity against both Gram-negative and Gram-positive bacteria. Conductive injectable self-healed hydrogels based on quaternized chitosan-g-polyaniline (QCSP) and benzaldehyde group functionalized poly(ethylene glycol)-co-poly(glycerol sebacate) (PEGS-FA) have also shown antibacterial, antioxidant and electroactive action for cutaneous wound healing [106]. Hydrogels based on chitosan and its derivatives have been broadly utilised as implant coatings because of their intrinsic non-toxic, osteoconductive properties, pH response, antimicrobial activity, biocompatibility and cell adhesive capacity [107, 108]. Novel hydrogel coatings produced by electrophoretic co-deposition of chitosan/alkynyl chitosan exhibited high antibacterial activity against *Escherichia coli* and *Staphylococcus aureus* [109] by the disk diffusion test [110, 111]. Antibacterial polymer coating tethered to the surface of medical implants and devices has attracted great interest in the last few decades for its ability to reduce implant-associated infections [112, 113]. Antimicrobial hydrogels of polyallylamine cross-linked with aldaric acid derivatives are very powerful weapons against a broad range of microorganisms: *Pseudomonas aeruginosa*, *E. coli*, *S. aureus* and *Candida albicans* [114]. Antibacterial ultrathin hydrogel films have been fabricated via a layer-by-layer (LbL) method and 'click' chemistry by Wang et al. [115]. This ultrathin hydrogel film consisted of poly[oligo(ethylene glycol) fumarate]-co-poly[dodecyl bis(2-hydroxyethyl) methyl ammonium fumarate] (POEGDMAM) containing multi-enes and poly[oligo(ethylene glycol) mercaptosuccinate] (POEGMS), and showed excellent antibacterial activity against both *S. aureus* and *E. coli*, due to the action of the ammonium groups with long alkyl chains present in the POEGDMAM. Doping antibiotics exogenously for eventual release in hydrogels has shown to be also an efficient antimicrobial strategy [116]. In these delivery systems, the antibacterial agent is released from the polymer matrix over time. Nevertheless, these systems present associated problems because the material's antibiotic release is eventually exhausted and the remaining polymer matrix may become a substrate for bacterial biofilm colonisation, which can become a concerning health threat. In these cases, secondary surgeries are carried out in order to remove these empty depots to prevent this type of infection risks. This second surgery can be avoided if a biodegradable hydrogel drug delivery system with a degradation rate linked via covalent incorporation of vancomycin in the hydrogel backbone [117]. Some antimicrobial fillers and/or agents need to be usually incorporated through physical blending in order to produce antimicrobial materials because most hydrogels themselves do not possess any antimicrobial action [118]. In this regard, graphene has emerged as a promising wide-spectrum antimicrobial nanomaterial, with unknown bacterial resistance so far and tolerable cytotoxic effect on mammalian cells. Through physical damage by direct contact of the sharp nanosheets' edges with bacterial membranes, graphene produces destructive extraction of lipid molecules. GO has also demonstrated to possess high antimicrobial activity against bacterial pathogens such as *Pseudomonas syringae* and *Xanthomonas campestris* pv. *undulosa*, and fungal pathogens such as *Fusarium graminearum* and *Fusarium oxysporum*. Thus, antimicrobial and biocompatible graphene-based nanocomposites have found application in a broad range of biomedical applications such as wound dressing [119], and silver coated medicinal devices, such as nanogels, nanolotions, etc. In the field of nanotechnology, silver nanoparticles (AgNPs) are also well-known for their antimicrobial activity and thus have been used to develop antimicrobial hydrogels with diverse biomedical applications such as silver-based dressings and silver coated medicinal devices, such as nanogels, nanolotions, etc. [120]. Microbial infections are frequent and are very undesired occurrences that may have occurred after orthopaedic procedures. Thus, medicated hydrogels of hyaluronic acid derivatives have been proposed [121] to address this problem. Nevertheless, there is a growing worldwide concern caused by the exponential increase in antibiotic resistance, which has open new alternative approaches such as the incorporation of AgNPs [122] or AgNPs combined with graphene [123] into hydrogels. Thus, an optimal mass ratio (5:1) of AgNPs with graphene showed excellent biocompatibility, high swelling ratio, good extensibility and much higher antimicrobial activity than other hydrogels. In addition, in vivo experiments performed in rats demonstrated that these nanocomposite hydrogels can accelerate the healing rate of artificial wounds. Similarly, acrylic acid (AA) grafted onto poly(ethylene terephthalate) (PET) film through gamma-ray-induced graft copolymerization with silver nanoparticles on the surface showed strong and stable antibacterial activity [124].

In the field of implanted biomedical devices, it is important to modify the hydrogel surface to achieve antifouling properties that make it resistant to protein adsorption and cell adhesion. Thus, for example, the antifouling activity of poly(2-hydroxyethyl methacrylate-co-methyl methacrylate) hydrogels was enhanced by surface grafting of a brush of poly(oligoethylene glycol methyl ether acrylate) [poly(OEGA)] [125]. Copolymerisation of non-fouling zwitterionic carboxybetaine methacrylamide (CBMAA-3) and 2-hydroxyethyl methacrylate (HEMA) in the presence of uniformly dispersed clay nanoparticles (Laponite XLG) in water by UV radiation has produced novel antifouling highly wettable hydrogels with superior mechanical performance and self-healing capacity [126]. In this field, it is highly desirable to produce hydrogels bearing antifouling properties and biocompatibility in order to prolong the lifetime of implanted materials, switchable antimicrobial property to eliminate infection and inflammation and outstanding mechanical performance to avoid the failure of the implanted material. To address these points, derivatives of zwitterionic carboxybetaine were developed with hydroxyl group(s), which can switch between the lactone form (antimicrobial) and the zwitterionic form (anti-fouling) [127]. Besides, the intramolecular hydrogen bonds enhance the mechanical property of the zwitterionic hydrogel, making it a viable material for coatings. However, the current increasing rates reported by the World Health Organisation of antibiotic resistance in pathogenic microbes are becoming a global health threat. Therefore, more resources and efforts must be done worldwide in order to develop new antimicrobial approaches such as antimicrobial hydrogels, because they have shown to be very effective in preventing and treating clinically relevant multidrug-resistant pathogens.

7. Porosity

The development of new advanced porous polymers has received much interest due to their potential to combine the properties of porous materials and polymers [128]. Many industrial fields such as gas storage and separation materials [129, 130], control drug delivery [131], catalysts [132], supports for electrochemical sensing [133], low-dielectric constant materials [134], packing materials for chromatography [135], and engineered three-dimensional porous matrices (*scaffolds*) for tissue regeneration in regenerative medicine [5, 45, 136, 137] require porous polymers. These advanced applications have driven much effort on developing new reliable techniques for fabricating porous polymers with specific pore architectures in the last few decades. Tissue engineering constitutes a promising alternative for tissue regeneration and even be able to bioengineer whole organ in the near future. Therefore, the development of new *scaffolds* has become a hot topic in biomedical research. Hydrogels have been proposed as leading candidates for engineered tissue *scaffolds* because of their good biocompatibility and close similarities to native extracellular matrix. Nevertheless, precise design of hydrogel properties, such as high and interconnected porosity suitable for specific cells, remains a challenge. Traditional methods for bulk porosity formation in hydrogels have demonstrated success for tissue engineering purposes. Nevertheless, some difficult issues related to direct cell encapsulation often occur. Thus, advanced technologies have shown to be able to produce hydrogels with suitable morphologies and function for tissue engineering applications [138]. In addition to the physicochemical properties and mechanical performance of the *scaffold* materials, the degree of interconnection and pore geometry, which depends on the tissue to be regenerate, plays a major role in these biomedical applications. Therefore, several methods and techniques have been developed so far to produce *scaffolds* with controlled morphology: gas foaming [139], fibre mesh sintering [140], solvent evaporation [141], polymerisation in the presence of a solvent [98, 142], porogen method [143, 144], freeze-drying [145, 146], electrospinning [147], 3D printing [148] and bioplotting [149], among many others. Thus, for instance, *scaffolds* of copolymerized hydrophobic ethyl acrylate (EA) and hydrophilic hydroxyethyl methacrylate comonomers with controlled hydrophilicity and interconnected morphology were prepared with a sintered template of controlled size PMMA microspheres [144]. On the other hand, porous *scaffolds* of gelatin with PHEMA have been synthesised by freeze-drying [146]. The HEMA content added in the initial mixtures before polymerisation modulated the porous architecture of the *scaffolds.* In addition, the covalently bound gelatin sequences significantly enhanced the biocompatibility of the PHEMA-based hydrogels. *Scaffolds* of high porosity can also be synthesised by the salt-leaching technique, utilising salts of NaCl or $(NH_4)_2SO_4$ as porogen agents [150] or with many other porogenic compounds such as sugar or ammonium oxalate crystals [151]. Carbon dioxide submitted to supercritical conditions and followed by rapid depressurization allows the fabrication of highly porous *scaffolds* [138]. Thus, *scaffolds* consisting of a blend of poly(ethyl methacrylate) and tetrahydrofurfuryl methacrylate with well interconnected and high porosity (greater than 85%) have been developed through this technique [152]. This gasification method has received significant attention in the past. Nevertheless, many scientists believe that the degree of pore interconnectivity is low. Electrospinning utilises a high-voltage DC power supply, infusion pumps and a syringe with a needle tip to produce fibres of varying diameters. For instance, a 3D aligned nanofiber-collagen type I hydrogel scaffold for controlled non-viral drug/gene delivery to direct axon regeneration in spinal cord injury treatment has been developed recently [153]. Nowadays, sophisticated biomedical devices designed by computer using patient-specific anatomical data can be easily fabricated by 3D printing. Thus, one-of-a-kind devices, surgical implants, *scaffolds* for tissue regeneration, and drug delivery systems have been fabricated by this relatively recent technique. Nevertheless, especially two technological limitations need still to be overcome: low choice of commercially available printable materials and very slow printing speed. Fused deposition modelling, selective laser sintering, stereolithography and 3D plotting/direct-write/bioprinting are also 3D printing techniques that are being progressively developed to suit tissue engineering applications. Bioprinting is the more advanced 3D printing technology because it consists of printing cells combined with custom 3D *scaffolds* for personalised regenerative medicine [148]. One of the important aspects in porous *scaffolds* is the mechanical resistance, which depends on the chemical and physical properties of the *scaffold* material and on its porosity. This aspect is even more relevant in hydrated porous polymer hydrogels at the body temperature in biomedical applications. Therefore, it is usually necessary to improve the mechanical properties of these porous hydrophilic materials by means of the mentioned reinforcing methods. Thus, for example, hybrid hydrogel nanocomposite *scaffolds* of silica/PHEA have shown enhanced mechanical performance in comparison with neat PHEA [45].

Another acrylic-based hydrogel, PHEMA, combined with cholesterol methacrylate (CHLMA) and lamina, has been synthesised in presence of ammonium oxalate crystals to produce *scaffolds* with interconnected pores that are able to promote cell-surface interaction [154]. The modification of PHEMA *scaffold*s with laminin-derived Ac-CGGASIKVAVS-OH peptide sequences has shown an enhanced effect to promote cell adhesion and neural differentiation. Cell adhesion and proliferation have also been improved in nanofiber *scaffolds* of poly(L-lactide) (PLLA) prepared by electrospinning by plasma treatment and simultaneously in situ grafting of hydrophilic acrylic acid to obtain PLLA-g-PAA [155]. Polysaccharide-based hydrogels have become increasingly proposed as matrices in soft tissue engineering because of their well-known cytocompatibility [68]. Thus, for instance, cross-linkable dextran methacrylates and hyaluronan methacrylate hydrogel matrices have been reported as leading candidates for soft tissue reconstruction and have shown that their in vitro degradation can be controlled by the polysaccharide morphology and cross-linking density. Besides, under in vitro conditions, these novel biomaterials had no toxic effects against fibroblasts and the use of composite gels improved cell adherence [156]. Therefore, proven advances have been reported in scaffold hydrogel design so far for tissue engineering applications. However, the need to develop viable *scaffolds* for clinical applications exists and new ways to find methods capable of providing suitable materials able to fulfil all the necessary requirements for different applications in regenerative medicine is a challenge.

8. Applications and future trends

Acrylate-based hydrogels are being widely investigated for their use in biomedical applications such as tissue engineering and systems for controlled delivery of biologically active agents. Hydrogels have been derivatized to enable crosslinking and photo-polymerisation, form electro-conducting networks and confer degradability to suit different biomedical applications. The pioneering work by Hubbel et al. [157] on bioerodible hydrogels based on copolymers of polylactic acid (PLA), copolymers of alpha hydroxyacids and polyethylene glycol (PEG) with acrylate end groups facilitated new tools for controlled release formulations and delivery platforms. These systems could be tuned to vary the degradation times from very short to long periods by changing composition of the hydrophobic ester block. Rational design using pH and thermoresponsive polymers have also been reported for drug delivery applications. A novel pH-responsive hydrogel based on carboxymethyl cellulose/2-hydroxyethyl acrylate was synthesised [158] as a transdermal delivery system for naringenin. The polymer exhibited different swelling ratio at different pH with Fickian diffusion characteristics, while mechanical properties could be controlled by varying cross-linking density and grafting. Photo-cross-linked hydrogels have also been used for the controlled release of various hydrophobic and hydrophilic drugs, including hydrogels based on methacrylate-terminated PEG [159] and PEG-poly(ε-caprolactone) multiblock copolymers [160]. Another family of hydrogels that is being explored for possible applications is thiol-acrylate hydrogels that were tailored to vary degradation rates and enhance cell viability particularly for cranial defects [161, 162]. A study of the multiscale modelling of the reaction kinetics of these thiol-acrylates with the mechanical properties revealed that the stiffness was related to light intensity, concentration of thiol groups and pH, thereby making it possible to tune the hydrogel properties. More recently, the synthesis and characterisation of novel acrylate-terminated amide-linked PEG-PDLA and PEG-PLLA star block copolymers were reported by Buwalda et al. [163]. Using PEG-PLA star block copolymer hydrogels through physical cross-linking and combining with photopolymerization, in situ gelation was achieved. Selected copolymers were initially physically cross-linked by stereocomplexation that on subsequent photopolymerization generated robust and stable hydrogels. This system was injectable and the hydrogels that were formed in situ were stable over periods of time that would be appropriate for long-term drug delivery.

9. Conclusions

Due to their versatile properties, acrylic-based hydrogels are currently used or proposed for a wide range of biomedical and bioengineering fields such as tissue engineering, orthopaedics, ophthalmology and dental materials. However, the number of applications of this type of biomaterials could be significantly increased if their properties would be improved, such as their mechanical strength in the hydrated state at the body temperature. In this regard, acrylic-based biomaterials have been developed as interpenetrating polymer networks, composites with incorporated particles such as fibres and/or advanced nanomaterials such as graphene, graphene oxide and carbon nanofibers, among many others, in order to improve their mechanical performance, electricity, thermal behaviour/degradation, water sorption/diffusion, biological interaction, antimicrobial activity and porosity values, required for certain applications. Furthermore, acrylic-based hydrogel scaffolds have been produced with the suitable morphology of pores and porosity for tissue regeneration by following developed protocols and sophisticated techniques. Nevertheless, in spite of the impressive advancements achieved so far, which are presented in this book chapter, many challenges still remain to design acrylate hydrogels with specific polymers for advanced applications in biomaterial science.

Acknowledgements

This work was supported by the 2019-231-003UCV grant from the Universidad Católica de Valencia San Vicente Mártir.

Author details

Ángel Serrano-Aroca[1]* and Sanjukta Deb[2]

1 Biomaterials and Bioengineering Lab, Centro de Investigación Traslacional San Alberto Magno, Universidad Católica de Valencia San Vicente Mártir, Valencia, Spain

2 Faculty of Dentistry, Oral and Craniofacial Sciences, King's College London, Guy's Hospital, London, UK

*Address all correspondence to: angel.serrano@ucv.es

References

[1] Buwalda SJ, Vermonden T, Hennink WE. Hydrogels for therapeutic delivery: Current developments and future directions. Biomacromolecules. 2017;**18**(2):316-330

[2] Chirila T, Harkin D. Biomaterials and Regenerative Medicine in Ophthalmology 2nd ed. Amsterdam, The Netherlands: Elsevier; 2016

[3] Shalaby S, Nagatomi S, Peniston S. Polymeric Biomaterials for Articulating Joint Repair and Total Joint Replacement. Polymers for Dental and Orthopedic Applications: CRC Press; 2007

[4] Kamoun EA, Kenawy E-RS, Chen X. A review on polymeric hydrogel membranes for wound dressing applications: PVA-based hydrogel dressings. Journal of Advanced Research. 2017;**8**(3):217-233

[5] Van Blitterswijk C, De Boer J. Tissue engineering. Oxford, United Kingdom: Academic Press; 2014

[6] Ratner BD, Hoffman AS, Schoen FJ, Lemons JE. Biomaterials Science: An Introduction to Materials in Medicine. Toronto, Canada: Academic Press; 2012

[7] Serrano-Aroca Á. Enhancement of Hydrogels' Properties for Biomedical Applications: Latest Achievements. In: Hydrogels. InTech; 2018. pp. 91-120

[8] Yue K, Trujillo-de Santiago G, Alvarez MM, Tamayol A, Annabi N, Khademhosseini A. Synthesis, properties, and biomedical applications of gelatin methacryloyl (GelMA) hydrogels. Biomaterials. 2015;**73**:254-271

[9] Stoy V, Climent C. In Hydrogels: Speciality Plastics for Biomedical and Pharmaceutical Applications. Basel: Technomic Publishers; 1996

[10] Serrano-Aroca Á, Monleón-Pradas M, Gómez-Ribelles JL. Effect of crosslinking on porous poly(methyl methacrylate) produced by phase separation. Colloid & Polymer Science. 2008;**286**(2):209-216

[11] Ramaraj B, Radhakrishnan G. Modification of the dynamic swelling behaviour of poly(2-hydroxyethyl methacrylate) hydrogels in water through interpenetrating polymer networks (IPNs). Polymer (Guildf). 1994;**35**(10):2167-2173

[12] Serrano-Aroca Á, Monleón-Pradas M, Gómez-Ribelles JL. Plasma-induced polymerisation of hydrophilic coatings onto macroporous hydrophobic scaffolds. Polymer (Guildf). 2007;**48**(7):2071-2078

[13] Serrano-Aroca Á, Gómez-Ribelles JL, Monleón-Pradas M, Vidaurre-Garayo A, Suay-Antón J. Characterisation of macroporous poly(methyl methacrylate) coated with plasma-polymerised poly (2-hydroxyethyl acrylate). European Polymer Journal. 2007;**43**(10):4552-4564

[14] Serrano-Aroca Á, Monleón-Pradas M, Gómez-Ribelles JL, Rault J. Thermal analysis of water in reinforced plasma-polymerised poly(2-hydroxyethyl acrylate) hydrogels. European Polymer Journal. 2015;**72**:523-534

[15] Gilbert JL, Ney DS, Lautenschlager EP. Self-reinforced composite poly (methyl methacrylate): Static and fatigue properties. Biomaterials. 1995;**16**(14):1043-1055

[16] Amerio E, Fabbri P, Malucelli G, Messori M, Sangermano M, Taurino R. Scratch resistance of nano-silica reinforced acrylic coatings. Progress in Organic Coatings. 2008;**62**(2):129-133

[17] Tahriri M, Del Monico M, Moghanian A, Tavakkoli Yaraki M, Torres R, Yadegari A, et al. Graphene and its derivatives: Opportunities and challenges in dentistry. Materials

Science and Engineering C. 2019;**102**:171-185

[18] Kugler S, Kowalczyk K, Spychaj T. Progress in organic coatings hybrid carbon nanotubes / graphene modified acrylic coats. Progress in Organic Coatings. 2015;**85**:1-7

[19] Cha C, Shin SR, Gao X, Annabi N, Dokmeci MR, Tang XS, et al. Controlling mechanical properties of cell-laden hydrogels by covalent incorporation of graphene oxide. Small. 2014;**10**(3):514-523

[20] Faghihi S, Gheysour M, Karimi A, Salarian R. Fabrication and mechanical characterization of graphene oxide-reinforced poly (acrylic acid)/gelatin composite hydrogels. Journal of Applied Physics. 2014;**115**:083513

[21] Sánchez-Correa F, Vidaurre-Agut C, Serrano-Aroca A, Campillo-Fernández AJ. Poly(2-hydroxyethyl acrylate) hydrogels reinforced with graphene oxide: Remarkable improvement of water diffusion and mechanical properties. Journal of Applied Polymer Science. 2017;**135**(15):46158

[22] Ha H, Shanmuganathan K, Ellison CJ. Mechanically stable thermally crosslinked poly(acrylic acid)/reduced graphene oxide aerogels. ACS Applied Materials & Interfaces. 2015;**7**(11):6220-6229

[23] Gong JP, Katsuyama Y, Kurokawa T, Osada Y. Double-network hydrogels with extremely high mechanical strength. Advanced Materials. 2003;**15**(14):1155-1158

[24] Sperling LH, Mishra V. The current status of interpenetrating polymer networks. Polymers for Advanced Technologies. 1996;**7**(4):197-208

[25] Maity J, Ray SK. Enhanced adsorption of methyl violet and Congo red by using semi and full IPN of polymethacrylic acid and chitosan. Carbohydrate Polymers. 2014;**104**(1):8-16

[26] Merlin DL, Sivasankar B. Synthesis and characterization of semi-interpenetrating polymer networks using biocompatible polyurethane and acrylamide monomer. European Polymer Journal. 2009;**45**(1):165-170

[27] Wang M, Pramoda KP, Goh SH. Mechanical behavior of pseudo-semi-interpenetrating polymer networks based on double-C 60 -end-capped poly (ethylene oxide) and poly (methyl methacrylate). Chemistry of Materials. 2004;**16**:3452-3456

[28] Xu LQ, Yao F, Fu GD, Kang ET. Interpenetrating network hydrogels via simultaneous "click chemistry" and atom transfer radical polymerization. Biomacromolecules. 2010;**11**(7):1810-1817

[29] Burugapalli K, Koul V, Dinda AK. Effect of composition of inter-penetrating polymer network hydrogels based on poly(acrylic acid) and gelatin on tissue response: A quantitative in vivo study. Journal of Biomedical Materials Research. Part A. 2004;**68**(2):210-218

[30] Shams Es-haghi S, Weiss RA. Finite strain damage-elastoplasticity in double-network hydrogels. Polym (United Kingdom). 2016;**103**:277-287

[31] Dragan ES. Design and applications of interpenetrating polymer network hydrogels. A review. Chemical Engineering Journal. 2014;**243**:572-590

[32] Lee J, Kim S, Kim S, Lee Y, Lee K. Synthesis and characteristics of interpenetrating polymer network hydrogel composed of chitosan and poly (acrylic acid). Polymer (Guildf). 1998;**73**(1):113-120

[33] Bajpai AK, Shukla SK, Bhanu S, Kankane S. Responsive polymers in controlled drug delivery. Progress in Polymer Science. 2008;**33**(11):1088-1118

[34] Zhao Y, Kang J, Tan T. Salt-, pH- and temperature-responsive semi-interpenetrating polymer network hydrogel based on poly(aspartic acid) and poly(acrylic acid). Polymer (Guildf). 2006;**47**(22):7702-7710

[35] Richter A, Paschew G, Klatt S, Lienig J, Arndt K-F, Adler H-JP. Review on hydrogel-based pH sensors and microsensors. Sensors. 2008;**8**(1):561-581

[36] DeKosky BJ, Dormer NH, Ingavle GC, Roatch CH, Lomakin J, Detamore MS, et al. Hierarchically designed Agarose and poly(ethylene glycol) interpenetrating network hydrogels for cartilage tissue engineering. Tissue Engineering. Part C, Methods. 2010;**16**(6):1533-1542

[37] Stumpel JE, Gil ER, Spoelstra AB, Bastiaansen CWM, Broer DJ, Schenning APHJ. Stimuli-responsive materials based on interpenetrating polymer liquid crystal hydrogels. Advanced Functional Materials. 2015;**25**(22):3314-3320

[38] Groover MP. Fundamentals of Modern Manufacturing_ Materials, Processes, and Systems 4th ed. New York, USA: John Wiley & Sons; 2010

[39] Young CD, Wu JR, Tsou TL. High-strength, ultra-thin and fiber-reinforced pHEMA artificial skin. Biomaterials. 1998;**19**(19):1745-1752

[40] Saheb DN, Jog JP. Natural fiber polymer composites: A review. Advances in Polymer Technology. 1999;**18**(4):351-363

[41] Ku H, Wang H, Pattarachaiyakoop N, Trada M. A review on the tensile properties of natural fiber reinforced polymer composites. Composites. Part B, Engineering. 2011;**42**(4):856-873

[42] Chen C, Li D, Hu Q, Wang R. Properties of polymethyl methacrylate-based nanocomposites: Reinforced with ultra-long chitin nanofiber extracted from crab shells. Materials and Design. 2014;**56**:1049-1056

[43] Kokubo T. Design of bioactive bone substitutes based on biomineralization process. Materials Science and Engineering: C. 2005;**25**(2):97-104

[44] Hench LL, West JK. The sol-gel process. Chemical Reviews. 1990;**90**(1): 33-72

[45] Rodríguez-Hernández JC, Serrano-Aroca Á, Gómez-Ribelles JL, Monleón-Pradas M. Three-dimensional nanocomposite scaffolds with ordered cylindrical orthogonal pores. Journal of Biomedical Materials Research. 2008;**84**(2):541-549

[46] Lin H-R, Ling M-H, Lin Y-J. High strength and low friction of a PAA-alginate-silica hydrogel as potential material for artificial soft tissues. Journal of Biomaterials Science. Polymer Edition. 2009;**20**(5-6):637-652

[47] Novoselov KS, Geim AK, Morozov SV, Jiang D, Zhang Y, Dubonos SV, et al. Electric field effect in atomically thin carbon films. Science. 2004;**306**(5696):666-669

[48] Geim AK, Novoselov KS. The rise of graphene. Nature Materials. 2007;**6**:183-191

[49] Balandin AA, Ghosh S, Bao W, Calizo I, Teweldebrhan D, Miao F, et al. Superior thermal conductivity of single-layer graphene. Nano Letters. 2008;**8**(3):902-907

[50] Lee C, Wei X, Kysar JW, Hone J. Measurement of the elastic properties and intrinsic strength of monolayer graphene. Science. 2008;**321**(5887):385-388

[51] Kalbacova M, Broz A, Kong J, Kalbac M. Graphene substrates promote adherence of human osteoblasts and mesenchymal stromal cells. Carbon New York. 2010;**48**(15):4323-4329

[52] Shen J, Yan B, Li T, Long Y, Li N, Ye M. Study on graphene-oxide-based polyacrylamide composite hydrogels. Composites. Part A, Applied Science and Manufacturing. 2012;**43**(9):1476-1481

[53] Shen J, Yan B, Li T, Long Y, Li N, Ye M. Mechanical, thermal and swelling properties of poly(acrylic acid)-graphene oxide composite hydrogels. Soft Matter. 2012;**8**(6):1831-1836

[54] Huang Y, Zeng M, Ren J, Wang J, Fan L, Xu Q. Preparation and swelling properties of graphene oxide/poly(acrylic acid-co-acrylamide) super-absorbent hydrogel nanocomposites. Colloids and Surfaces A: Physicochemical and Engineering. 2012;**401**:97-106

[55] Serrano-Aroca Á, Iskandar L, Deb S. Green synthetic routes to alginate-graphene oxide composite hydrogels with enhanced physical properties for bioengineering applications. European Polymer Journal. 2018;**103**:198-206. DOI: 10.1016/j.eurpolymj.2018.04.015

[56] Martí M, Frígols B, Salesa B, Serrano-Aroca Á. Calcium alginate/graphene oxide films: Reinforced composites able to prevent Staphylococcus aureus and methicillin-resistant Staphylococcus epidermidis infections with no cytotoxicity for human keratinocyte HaCaT cells. European Polymer Journal. 2019;**110**:14-21

[57] Frígols B, Martí M, Salesa B, Hernández-Oliver C, Aarstad O, Ulset AST, et al. Graphene oxide in zinc alginate films: Antibacterial activity, cytotoxicity, zinc release, water sorption/diffusion, wettability and opacity. Ng DYW, editor. PLoS One. 2019;**14**(3):e0212819

[58] Xu Y, Wu Q, Sun Y, Bai H, Shi G. Three-dimensional self-assembly of graphene oxide and DNA into multifunctional hydrogels. ACS Nano. 2010;**4**(12):7358-7362

[59] Serrano-Aroca Á, Deb S. Synthesis of irregular graphene oxide tubes using green chemistry and their potential use as reinforcement materials for biomedical applications. PLoS One. 2017 Sep;**12**(9):e0185235

[60] Iijima S. Helical microtubules of graphitic carbon. Nature. 1991; **354**(6348):56-58

[61] Ahadian S, Ramón-Azcón J, Estili M, Liang X, Ostrovidov S, Shiku H, et al. Hybrid hydrogels containing vertically aligned carbon nanotubes with anisotropic electrical conductivity for muscle myofiber fabrication. Scientific Reports. 2014;**4**:4271

[62] Ogoshi T, Takashima Y, Yamaguchi H, Harada A. Chemically-responsive sol-gel transition of supramol single-walled carbon nanotube SWNT hydrogel made by hybrid of SWNTs & cyclodextrin. Journal of the American Chemical Society. 2007;**129**:4878-4879

[63] Satarkar NS, Johnson D, Marrs B, Andrews R, Poh C, Gharaibeh B, et al. Hydrogel-MWCNT nanocomposites: Synthesis, characterization, and heating with radiofrequency fields. Journal of Applied Polymer Science. 2010;**117**(3):1813-1819

[64] Bhattacharyya S, Guillot S, Dabboue H, Tranchant JF, Salvetat JP. Carbon nanotubes as structural nanofibers for hyaluronic acid hydrogel scaffolds. Biomacromolecules. 2008;**9**(2):505-509

[65] Meng X, Stout DA, Sun L, Beingessner RL, Fenniri H, Webster TJ. Novel injectable biomimetic hydrogels with carbon nanofibers and self assembled rosette nanotubes for

myocardial applications. Journal of Biomedical Materials Research Part A. 2013;**101A**(4):1095-1102

[66] Salesa B, Martí M, Frígols B, Serrano-Aroca Á. Carbon nanofibers in pure form and in calcium alginate composites films: New cost-effective antibacterial biomaterials against the life-threatening multidrug-resistant Staphylococcus epidermidis. Polymers (Basel). 2019;**11**(3):453 Available from: https://www.mdpi.com/2073-4360/11/3/453

[67] Llorens-Gámez M, Salesa B, Serrano-Aroca Á. Physical and biological properties of alginate/carbon nanofibers hydrogel films. International Journal of Biological Macromolecules. 2020;**151**:499-507. DOI: 10.1016/j.ijbiomac.2020.02.213

[68] Salesa B, Llorens-Gámez M, Serrano-Aroca Á. Study of 1D and 2D carbon nanomaterial in alginate films. Nanomaterials. 2020;**10**(2):206

[69] Iwamoto S, Nakagaito AN, Yano H, Nogi M. Optically transparent composites reinforced with plant fiber-based nanofibers. Applied Physics A: Materials Science & Processing. 2005;**81**(6):1109-1112

[70] Zhou C, Wu Q. A novel polyacrylamide nanocomposite hydrogel reinforced with natural chitosan nanofibers. Colloids Surfaces B Biointerfaces. 2011;**84**(1):155-162

[71] Kokabi M, Sirousazar M, Hassan ZM. PVA-clay nanocomposite hydrogels for wound dressing. European Polymer Journal. 2007;**43**(3):773-781

[72] Meneghetti P, Qutubuddin S. Synthesis, thermal properties and applications of polymer-clay nanocomposites. Thermochimica Acta. 2006;**442**(1-2):74-77

[73] Zhao M, Forrester JV, McCaig CD. A small, physiological electric field orients cell division. Proceedings of the National Academy of Sciences of the United States of America. 1999;**96**:4942-4946

[74] Yao L, Shanley L, Mccaig C, Zhao M. Small applied electric fields guide migration of hippocampal neurons. Journal of Cellular Physiology. 2008 Aug;**216**(2):527-535

[75] Woo DG, Shim MS, Park JS, Yang HN, Lee DR, Park KH. The effect of electrical stimulation on the differentiation of hESCs adhered onto fibronectin-coated gold nanoparticles. Biomaterials. 2009;**30**(29):5631-5638

[76] Sisken BF, Walker J, Orgel M. Prospects on clinical applications of electrical stimulation for nerve regeneration. Journal of Cellular Biochemistry. 1993;**51**(4):404-409

[77] Zhao M. Electrical fields in wound healing-an overriding signal that directs cell migration. Seminars in Cell & Developmental Biology. 2009;**20**(6):674-682

[78] Kirson ED, Dbalý V, Tovaryš F, Vymazal J, Soustiel JF, Itzhaki A, et al. Alternating electric fields arrest cell proliferation in animal tumor models and human brain tumors. Proceedings of the National Academy of Sciences. 2007;**104**(24):10152-10157

[79] Fabbri P, Valentini L, Bittolo Bon S, Foix D, Pasquali L, Montecchi M, et al. In-situ graphene oxide reduction during UV-photopolymerization of graphene oxide/acrylic resins mixtures. Polymer (Guildf). 2012;**53**(26):6039-6044

[80] Huang YL, Tien HW, Ma CC, Yang SY, Wu SY, Liu HY, et al. Effect of extended polymer chains on properties of transparent graphene nanosheets conductive film. Journal of Materials Chemistry. 2011;**21**(45):18236

[81] Huang P, Chen W, Yan L. An inorganic-organic double network hydrogel of graphene and polymer. Nanoscale. 2013;**5**(13):6034-6039

[82] Rao CNR, Satishkumar BC, Govindaraj A. Nanotubes. ChemPhysChem. 2001;**2**:78-105

[83] Wang ZG, Wang Y, Xu H, Li G, Xu ZK. Carbon nanotube-filled Nanofibrous membranes electrospun from poly(acrylonitrile- co -acrylic acid) for glucose biosensor. Journal of Physical Chemistry C. 2009;**113**(7):2955-2960

[84] Ramón-Azcón J, Ahadian S, Estili M, Liang X, Ostrovidov S, Kaji H, et al. Dielectrophoretically aligned carbon nanotubes to control electrical and mechanical properties of hydrogels to fabricate contractile muscle myofibers. Advanced Materials. 2013;**25**(29):4028-4034

[85] Huang ZH, Qiu KY. The effects of interactions on the properties of acrylic polymers/silica hybrid materials prepared by the in situ sol-gel process. Polymer (Guildf). 1997;**38**(3):521-526

[86] Ramanathan T, Abdala AA, Stankovich S, Dikin DA, Herrera-Alonso M, Piner RD, et al. Functionalized graphene sheets for polymer nanocomposites. Nature Nanotechnology. 2008;**3**(6):327-331

[87] Fuhrer R, Athanassiou EK, Luechinger NA, Stark WJ. Crosslinking metal nanoparticles into the polymer backbone of hydrogels enables preparation of soft, magnetic field-driven actuators with muscle-like flexibility. Small. 2009;**5**(3):383-388

[88] Dash R, Foston M, Ragauskas AJ. Improving the mechanical and thermal properties of gelatin hydrogels cross-linked by cellulose nanowhiskers. Carbohydrate Polymers. 2013;**91**(2): 638-645

[89] Roorda WE, Bouwstra JA, de Vries MA, Junginger HE. Thermal analysis of water in p(HEMA) hydrogels. Biomaterials. 1988;**9**(6):494-499

[90] Serrano-Aroca Á, Campillo-Fernández AJ, Gómez-Ribelles JL, Monleón-Pradas M, Gallego-Ferrer G, Pissis P. Porous poly(2-hydroxyethyl acrylate) hydrogels prepared by radical polymerisation with methanol as diluent. Polymer (Guildf). 2004;**45**(26):8949-8955

[91] Clayton AB, Chirila TV, Lou X. Hydrophilic sponges based on 2-Hydroxyethyl methacrylate. V. Effect of crosslinking agent reactivity on mechanical properties. Polymer International. 1997;**44**:201-207

[92] Monleón-Pradas M, Gómez-Ribelles JL, Serrano-Aroca Á, Gallego-Ferrer G, Suay-Antón J, Pissis P. Interaction between water and polymer chains in poly(hydroxyethyl acrylate) hydrogels. Colloid & Polymer Science. 2001;**279**(4):323-330

[93] Ahmed EM. Hydrogel: Preparation, characterization, and applications: A review. Journal of Advanced Research. 2015;**6**(2):105-121

[94] Tanaka T. Gels. Scientific American. 1981;**244**(1):124-136 138

[95] Serrano-Aroca Á, Ruiz-Pividal J-F, Llorens-Gámez M. Enhancement of water diffusion and compression performance of crosslinked alginate films with a minuscule amount of graphene oxide. Scientific Reports. 2017;**7**:11684. DOI: 10.1038/s41598-017-10260-x

[96] Crank J. The Mathematics of Diffusion Second Edition. Bristol, United Kingdo: Oxford University Press; 1975

[97] Wang X, Zhang Y, Hao C, Dai X, Zhu F, Ge C. Ultrasonic synthesis and properties of a sodium lignosulfonate–grafted poly(acrylic acid-co-acryl

amide) composite super absorbent polymer. New Journal of Chemistry. 2014;**38**(12):6057-6063

[98] Monleón-Pradas M, Gómez-Ribelles JL, Serrano-Aroca Á, Gallego Ferrer G, Suay Antón J, Pissis P. Porous poly (2-hydroxyethyl acrylate) hydrogels. Polymer (Guildf). 2001;**42**(10):4667-4674

[99] Gehrke SH, Biren D, Hopkins JJ. Evidence for Fickian water transport in initially glassy poly(2-hydroxyethyl methacrylate). Journal of Biomaterials Science. Polymer Edition. 1995;**6**(4):375-390

[100] Wang J, Wu W. Swelling behaviors, tensile properties and thermodynamic studies of water sorption of 2-hydroxyethyl methacrylate/epoxy methacrylate copolymeric hydrogels. European Polymer Journal. 2005;**41**(5):1143-1151

[101] Kavitha T, Kang IK, Park SY. Poly(acrylic acid)-grafted graphene oxide as an intracellular protein carrier. Langmuir. 2014;**30**(1):402-409

[102] Krishna-Rao KSV, Ha CS. PH sensitive hydrogels based on acryl amides and their swelling and diffusion characteristics with drug delivery behavior. Polymer Bulletin. 2009;**62**(2): 167-181

[103] Zhang XZ, Yang YY, Wang FJ, Chung TS. Thermosensitive poly(N-isopropylacrylamide-co-acrylic acid) hydrogels with expanded network structures and improved oscillating swelling-deswelling properties. Langmuir. 2002;**18**(6):2013-2018

[104] Shi L, Chen J, Teng L, Wang L, Zhu G, Liu S, et al. The antibacterial applications of Graphene and its derivatives. Small. 2016;**12**(31):4165-4184

[105] Giano MC, Ibrahim Z, Medina SH, Sarhane KA, Christensen JM, Yamada Y, et al. Injectable bioadhesive hydrogels with innate antibacterial properties. Nature Communications. 2014;**5**(May): 1-9

[106] Zhao X, Wu H, Guo B, Dong R, Qiu Y, Ma PX. Antibacterial anti-oxidant electroactive injectable hydrogel as self-healing wound dressing with hemostasis and adhesiveness for cutaneous wound healing. Biomaterials. 2017;**122**:34-47

[107] Jayakumar R, Prabaharan M, Nair SV, Tamura H. Novel chitin and chitosan nanofibers in biomedical applications. Biotechnology Advances. 2010;**28**:142-150

[108] Jayakumar R, Prabaharan M, Sudheesh Kumar PT, Nair SV, Tamura H. Biomaterials based on chitin and chitosan in wound dressing applications. Biotechnology Advances. 2011;**29**:322-337

[109] Ding F, Nie Z, Deng H, Xiao L, Du Y, Shi X. Antibacterial hydrogel coating by electrophoretic co-deposition of chitosan/alkynyl chitosan. Carbohydrate Polymers. 2013;**98**(2):1547-1552

[110] Bauer AW, Kirby WMM, Sherris JC, Turck AM. Antibiotic susceptibility testing by a standardized single disk method. American Journal of Clinical Pathology. 1966;**45**:493-496

[111] Martí M, Frígols B, Serrano-Aroca Á. Antimicrobial characterization of advanced materials for bioengineering applications. Journal of Visualized Experiments. 2018;**138**:e57710

[112] Zhou C, Li P, Qi X, Sharif ARM, Poon YF, Cao Y, et al. A photopolymerized antimicrobial hydrogel coating derived from epsilon-poly-l-lysine. Biomaterials. 2011;**32**(11):2704-2712

[113] Li P, Poon YF, Li W, Zhu H-Y, Yeap SH, Cao Y, et al. A polycationic

antimicrobial and biocompatible hydrogel with microbe membrane suctioning ability. Nature Materials. 2011;**10**(2):149-156

[114] Andrews MA, Figuly GD, Chapman JS, Hunt TW, Glunt CD, Rivenbark JA, et al. Antimicrobial hydrogels formed by crosslinking polyallylamine with aldaric acid derivatives. Journal of Applied Polymer Science. 2011;**119**(6):3244-3252

[115] Wang H, Zha G, Du H, Gao L, Li X, Shen Z, et al. Facile fabrication of ultrathin antibacterial hydrogel films via layer-by-layer "click" chemistry. Polymer Chemistry. 2014;**5**(22): 6489-6494

[116] Smith AW. Biofilms and antibiotic therapy: Is there a role for combating bacterial resistance by the use of novel drug delivery systems? Advanced Drug Delivery Reviews. 2005;**57**:1539-1550

[117] Lakes AL, Peyyala R, Ebersole JL, Puleo DA, Hilt JZ, Dziubla TD. Synthesis and characterization of an antibacterial hydrogel containing covalently bound vancomycin. Biomacromolecules. 2014;**15**(8): 3009-3018

[118] He J, Söderling E, Lassila LVJ, Vallittu PK. Incorporation of an antibacterial and radiopaque monomer in to dental resin system. Dental Materials. 2012;**28**(8):e110-e117

[119] Ji H, Sun H, Qu X. Antibacterial applications of graphene-based nanomaterials: Recent achievements and challenges. Advanced Drug Delivery Reviews. 2016;**105**:176-189

[120] Rai M, Yadav A, Gade A. Silver nanoparticles as a new generation of antimicrobials. Biotechnology Advances. 2009;**27**:76-83

[121] Pitarresi G, Palumbo FS, Calascibetta F, Fiorica C, Di Stefano M, Giammona G. Medicated hydrogels of hyaluronic acid derivatives for use in orthopedic field. International Journal of Pharmaceutics. 2013;**449**(1-2):84-94

[122] González-Sánchez MI, Perni S, Tommasi G, Morris NG, Hawkins K, López-Cabarcos E, et al. Silver nanoparticle based antibacterial methacrylate hydrogels potential for bone graft applications. Materials Science and Engineering: C. 2015;**50**:332-340

[123] Fan Z, Liu B, Wang J, Zhang S, Lin Q, Gong P, et al. A novel wound dressing based on Ag/graphene polymer hydrogel: Effectively kill bacteria and accelerate wound healing. Advanced Functional Materials. 2014;**24**(25):3933-3943

[124] Ping X, Wang M, Xuewu G. Surface modification of poly(ethylene terephthalate) (PET) film by gamma-ray induced grafting of poly(acrylic acid) and its application in antibacterial hybrid film. Radiation Physics and Chemistry. 2011;**80**(4):567-572

[125] Bozukova D, Pagnoulle C, De Pauw-Gillet MC, Ruth N, Jérôme R, Jérôme C. Imparting antifouling properties of poly(2-hydroxyethyl methacrylate) hydrogels by grafting poly(oligoethylene glycol methyl ether acrylate). Langmuir. 2008;**24**(13):6649-6658

[126] Kostina NY, Sharifi S, de los Santos Pereira A, Michálek J, Grijpma DW, Rodriguez-Emmenegger C. Novel antifouling self-healing poly (carboxybetaine methacrylamide-co-HEMA) nanocomposite hydrogels with superior mechanical properties. Journal of Materials Chemistry B. 2013;**1**(41):5644

[127] Cao B, Tang Q, Li L, Humble J, Wu H, Liu L, et al. Switchable antimicrobial and antifouling hydrogels with enhanced mechanical properties. Advanced Healthcare Materials. 2013;**2**(8):1096-1102

[128] Wu D, Xu F, Sun B, Fu R, He H, M atyjaszewski K. Design and preparation of porous polymers. Chemical Reviews. 2012;**112**:3959-4015

[129] Du N, Park HB, Robertson GP, Dal-Cin MM, Visser T, Scoles L, et al. Polymer nanosieve membranes for CO2-capture applications. Nature Materials. 2011;**10**(5):372-375

[130] Shetty D, Jahovic I, Raya J, Ravaux F, Jouiad M, Olsen J-C, et al. An ultra-absorbent alkyne-rich porous covalent polycalix[4]arene for water purification. Journal of Materials Chemistry A. 2017;**5**:62

[131] Kolanthai E, Dikeshwar Colon VS, Sindu PA, Chandra VS, Karthikeyan KR, Babu MS, et al. Effect of solvent; enhancing the wettability and engineering the porous structure of a calcium phosphate/agarose composite for drug delivery. RSC Advances. 2015;**5**(24):18301-18311

[132] Pulko I, Wall J, Krajnc P, Cameron NR. Ultra-high surface area functional porous polymers by emulsion templating and hypercrosslinking: Efficient nucleophilic catalyst supports. Chemistry: A European Journal. 2010;**16**(8):2350-2354

[133] Zhao C, Danish E, Cameron NR, Kataky R. Emulsion-templated porous materials (PolyHIPEs) for selective ion and molecular recognition and transport: Applications in electrochemical sensing. Journal of Materials Chemistry. 2007;**17**(23):2446

[134] Fu GD, Yuan Z, Kang ET, Neoh KG, Lai DM, Huan ACH. Nanoporous ultra-low-dielectric-constant fluoropolymer films via selective UV decomposition of poly(pentafluorostyrene)-block-poly(methyl methacrylate) copolymers prepared using atom transfer radical polymerization. Advanced Functional Materials. 2005;**15**(2):315-322

[135] Lv Y, Hughes TC, Hao X, Hart NK, Littler SW, Zhang X, et al. A novel route to prepare highly reactive and versatile chromatographic monoliths. Macromolecular Rapid Communications. 2010;**31**(20):1785-1790

[136] Murphy AR, Laslett A, O'Brien CM, Cameron NR. Scaffolds for 3D In vitro culture of neural lineage cells. Acta Biomaterialia. 2017;**54**:1-20

[137] Serrano-Aroca Á, Llorens-Gámez M. Dynamic mechanical analysis and water vapour sorption of highly porous poly(methyl methacrylate). Polymer (Guildf). 2017;**125**:58-65

[138] Annabi N, Nichol JW, Zhong X, Ji C, Koshy S, Khademhosseini A, et al. Controlling the porosity and micro-architecture of hydrogels for tissue engineering. Tissue Engineering. Part B, Reviews. 2010;**16**(4):371-383

[139] Arora KA, Lesser AJ, McCarthy TJ. Compressive behavior of microcellular polystyrene foams processed in supercritical carbon dioxide. Polymer Engineering and Science. 1998;**38**(12):2055-2062

[140] Thomson RC, Wake MC, Yaszemski MJ, Mikos AG. Biodegradable polymer scaffolds to regenerate organs. Advances in Polymer Science. 1995;**122**:247-277

[141] Andrianova GP, Pakhomov SI. Porous materials from crystallizable polyolefins produced by gel technology. Polymer Engineering and Science. 1997;**37**(8):1367-1380

[142] Serrano-Aroca Á, Monleón-Pradas M, Gómez-Ribelles JL. Macroporous poly(methyl methacrylate) produced by phase separation during polymerisation in solution. Colloid & Polymer Science. 2007;**285**(7):753-760

[143] Flynn L, Dalton PD, Shoichet MS. Fiber templating of poly(2-hydroxyethyl methacrylate) for neural tissue

engineering. Biomaterials. 2003;**24**(23): 4265-4272

[144] Brígido-Diego R, Pérez-Olmedilla M, Serrano-Aroca Á, Gómez-Ribelles JL, Monleón-Pradas M, Gallego-Ferrer G, et al. Acrylic scaffolds with interconnected spherical pores and controlled hydrophilicity for tissue engineering. Journal of Materials Science. Materials in Medicine. 2005;**40**(18):4881-4887

[145] Kang HW, Tabata Y, Ikada Y. Fabrication of porous gelatin scaffolds for tissue engineering. Biomaterials. 1999;**20**(14):1339-1344

[146] Dragusin DM, Van Vlierberghe S, Dubruel P, Dierick M, Van Hoorebeke L, Declercq HA, et al. Novel gelatin–PHEMA porous scaffolds for tissue engineering applications. Soft Matter. 2012;**8**(37):9589

[147] Agarwal S, Wendorff JH, Greiner A. Use of electrospinning technique for biomedical applications. Polymer (Guildf). 2008;**49**(26):5603-5621

[148] Chia HN, Wu BM. Recent advances in 3D printing of tissue engineering scaffolds. Journal of Biological Engineering. 2015;**9**(4):2-14

[149] Derby B. Printing and prototyping of tissues and scaffolds. Science. 2012;**338**(6109):921-926

[150] Horák D, Hlídková H, Hradil J, Lapčíková M, Šlouf M. Superporous poly(2-hydroxyethyl methacrylate) based scaffolds: Preparation and characterization. Polymer (Guildf). 2008;**49**(8):2046-2054

[151] Kubinová Š, Horák D, Kozubenko N, Vaněček V, Proks V, Price J, et al. The use of superporous Ac-CGGASIKVAVS-OH-modified PHEMA scaffolds to promote cell adhesion and the differentiation of human fetal neural precursors. Biomaterials. 2010;**31**(23):5966-5975

[152] Barry JJA, Silva MMCG, Cartmell SH, Guldberg RE, Scotchford CA, Howdle SM. Porous methacrylate tissue engineering scaffolds: Using carbon dioxide to control porosity and interconnectivity. Journal of Materials Science. 2006;**41**(13):4197-4204

[153] Nguyen LH, Gao M, Lin J, Wu W, Wang J, Chew SY. Three-dimensional aligned nanofibers-hydrogel scaffold for controlled non-viral drug/gene delivery to direct axon regeneration in spinal cord injury treatment. Scientific Reports. 2017;**7**(October 2016):42212

[154] Kubinová Š, Horák D, Syková E. Cholesterol-modified superporous poly(2-hydroxyethyl methacrylate) scaffolds for tissue engineering. Biomaterials. 2009;**30**(27):4601-4609

[155] Park K, Hyun JJ, Kim JJ, Ahn KD, Dong KH, Young MJ. Acrylic acid-grafted hydrophilic electrospun nanofibrous poly(L-lactic acid) scaffold. Macromolecular Research. 2006;**14**(5):552-558

[156] Möller S, Weisser J, Bischoff S, Schnabelrauch M. Dextran and hyaluronan methacrylate based hydrogels as matrices for soft tissue reconstruction. Biomolecular Engineering. 2007;**24**(5):496-504

[157] Sawhney AS, Pathak CP, Hubbell JA. Bioerodible hydrogels based on Photopolymerized poly(ethylene glycol)-co-poly (α-hydroxy acid) Diacrylate Macromers. Macromolecules. 1993;**26**(4):581-587

[158] Park SH, Shin HS, Park SN. A novel pH-responsive hydrogel based on carboxymethyl cellulose/2-hydroxyethyl acrylate for transdermal delivery of naringenin. Carbohydrate Polymers. 2018;**200**:341-352

[159] DiRamio JA, Kisaalita WS, Majetich GF, Shimkus JM. Poly(ethylene

glycol) methacrylate/dimethacrylate hydrogels for controlled release of hydrophobic drugs. Biotechnology Progress. 2005;**21**(4):1281-1288

[160] Lee WF, Cheng TS. Synthesis and drug-release behavior of porous biodegradable amphiphilic co-polymeric hydrogels. Journal of Biomaterials Science. Polymer Edition. 2009;**20**(14):2023-2037

[161] Emmakah AM, Arman HE, Bragg JC, Greene T, Alvarez MB, Childress PJ, et al. A fast-degrading thiol-acrylate based hydrogel for cranial regeneration. Biomedical Materials. 2017;**12**(2):025011

[162] Zhu H, Yang X, Genin GM, Lu TJ, Xu F, Lin M. The relationship between thiol-acrylate photopolymerization kinetics and hydrogel mechanics: An improved model incorporating photobleaching and thiol-Michael addition. Journal of the Mechanical Behavior of Biomedical Materials. 2018;**88**:160-169

[163] Buwalda SJ, Dijkstra PJ, Feijen J. In situ forming stereocomplexed and post-photocrosslinked acrylated star poly(ethylene glycol)-poly(lactide) hydrogels. European Polymer Journal. 2017;**94**:152-161